FIRESIDE

Simon and Schuster's

Guide to
MUSHROOMS

By Giovanni Pacioni

U.S. Editor: Gary Lincoff

A Fireside Book

Published by Simon and Schuster

New York

Copyright © 1981 by Arnoldo Mondadori Editore S.p.A., Milano
English language translation copyright © 1981 by Arnoldo Mondadori Editore S.p.A., Milano
All rights reserved
including the right of reproduction
in whole or in part in any form
A Fireside Book
Published by Simon and Schuster
A Division of Gulf & Western Corporation
Simon & Schuster Building
Rockefeller Center
1230 Avenue of the Americas
New York, New York 10020

Originally published in Italy under the title *Funghi* in 1980, by Arnoldo Mondadori Editore S.p.A., Milano

FIRESIDE and colophon are trademarks of Simon & Schuster

Manufactured by Officine Grafiche di Arnoldo Mondadori Editore, Verona
Printed in Italy

10   9   8   7   6   5   4   3   2   1
10   9   8   7   6   5   4   3   2   1 Pbk.

Library of Congress Cataloging in Publication Data
Main entry under title:
Simon and Schuster's guide to mushrooms.
    (A Fireside book)
    Translation of Funghi.
    Includes index.
    1. Mushrooms—Identification. I. Pacioni, Giovanni.
II. Lincoff, Gary.   III. Title: Guide
to mushrooms.
QK617.P23413   589.2   81-16605
ISBN 0-671-42798-9   AACR2
       0-671-42849-7 (Pbk.)

English translation by Simon Pleasance.
Illustrations by Andrea Corbella
Symbols and layout by Giorgio Seppi

CONTENTS

The illustrated section of this book has been divided into five parts: (1) fungi with cap and stipe and gills (**1–238**); (2) fungi with cap and stipe and pores or teeth, or hood-shaped (**239–288**); (3) bracket, crust, and jelly fungi (**289–324**); (4) bush- and club-shaped fungi (**325–352**); and (5) sphere-, star-, pear-, and cup-shaped fungi (**353–420**). The fungi are thus arranged, regardless of the systematic group to which they belong, on the basis of their morphological appearance, following the order proposed in the key (see page 58–59). The names of the species are faithful to current nomenclature, even though, in the keys, reference is often made to past genera which have now been considerably divided up, e.g. *Polyporus*. This is in order not to overload a section designed to focus on methods of observation, and not intended as an analytic treatise. In the remarks about edibility, the adjectives "edible," "mediocre," "fair," or "good" should always be taken to mean "after being correctly cooked!"

The English language is probably alone in attempting to make a clear distinction between mushrooms (edible), toadstools (inedible or poisonous mushrooms) and fungi (a term most commonly applied to the bracketlike growths we see on trees, but which also refers to molds, mildews, and other organisms in the kingdom which do not have the traditional stipe-and-cap shape of mushrooms). The term "fungi" will nevertheless be used in this text in the broad sense that includes both edible and poisonous mushrooms as well as bracket fungi, mildews, and molds.

KEY TO SYMBOLS

SPORE COLOR

Spore color white, whitish, yellow

pinkish to salmon

yellow-brown to rust-brown

dark purple-brown, violet, and black

EDIBILITY

CAUTION

READ WITH CARE

deadly

poisonous

inedible

good

excellent

TYPE OF FUNGUS IN RELATION TO SURROUNDING PLANT ENVIRONMENT

symbiotic

parasitic

saprophytic

FUNGI AND THE ENVIRONMENT

What are fungi? Fungi are organisms which have a nucleus, lack the pigment chlorophyll, originate from spores and reproduce both sexually and asexually, and whose normally filamentous (threadlike) and ramified (branching) somatic structures are surrounded by cellular walls containing cellulose or chitin, or both. This definition, quite correct from a scientific point of view, has the drawback of offering very little tangible information to most people who are not specialists in the field. It is therefore a useful idea to include an abbreviated definition which may not cover all the different species of fungi, but which will be much more intelligible to most readers: Fungi are the fruit of an organism which grows either in the ground or on an organic substratum that is either living or dead.

Both these definitions will become easier to grasp if we outline the life cycle of fungi which produce carpophores, i.e. the *fruit*, normally called *mushrooms*.

The seed that triggers off the birth of the fungus is microscopic in size and called a *spore*. The spore germinates and sends out a small tube which becomes enlarged at the tip, where very large numbers of cells are formed in quick succession. A filament is thus formed which becomes variously ramified, and each new tip of each ramification is a new growth center. The filament formed by successively arranged cells is called a *hypha,* and together the hyphae constitute the *mycelium.* By means of the sexual process the carpophore will form from the mycelium. Two hyphae of the same mycelium or, more commonly, of different mycelia, join together to form a new mycelium, hallmarked by the stimultaneous presence of two nuclei, one originating from each hypha. When the environment and nutritional conditions are favorable, the fruit-

The development cycle of a fungus. A spore germinates, producing a primary mycelium which comes into contact with the mycelium produced by another spore. Their union gives rise to the secondary mycelium which will form what is technically known as the carpophore (fruit- or fruiting body).

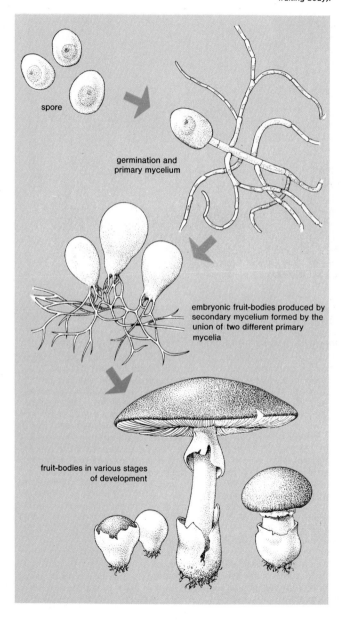

spore

germination and
primary mycelium

embryonic fruit-bodies produced by
secondary mycelium formed by the
union of two different primary
mycelia

fruit-bodies in various stages
of development

A particularly conspicuous mycelium which will eventually produce the fruit-bodies of Coprinus radians, *(230) a species of ink cap.*

body—or mushroom proper—will develop from this mycelium and produce the spores. Depending on the type of fungus in question, the mycelium can live on various substrates: in the ground, or on dead or living organic matter. In the ground, the mycelium feeds by decomposing the organic substances present, or by striking up a special "mutual-aid" relationship with green plants—i.e. trees, grasses, ferns, etc. The relationship between fungi and the roots of green plants is a symbiotic one known as *mycorrhiza* or *mycorrhizal symbiosis,* whereas the breaking down of dead organic matter or substances is called *saprophytism.* The term saprophyte also applies to those fungi which grow on dead wood, leaves, fruit, or excrement. When a fungus establishes itself on a living organism, and causes it damage, which may be fatal, we use the term *parasitism.* Parasitic fungi live on various parts of plants, or on leaves, stalks, roots, flowers, or fruits, on animals—especially insects—and on other fungi.

Mycorrhiza or mycorrhizal symbiosis
The term mycorrhiza describes a symbiotic relationship between certain fungi and the roots of almost all green plants. Green plants contain specific pigments called chlorophylls which, with the help of sunlight, enable them to synthesize organic substances essential for their nutrition: sugars or carbo-

hydrates. Because of this capacity, they are called *autotro-phic,* i.e. "self-feeding." The fungi, on the other hand, like many bacteria and the members of the animal kingdom, are *heterotrophic* as far as carbohydrates are concerned. That is, they have to take the carbohydrates they require from other organisms. Autotrophic plants, whether growing on land or in water, are therefore the basis of our food chain.

In the case of mycorrhizal symbiosis the fungus extracts excess reserve sugars——mainly starch——from the roots of plants. But because of the way the mycelium is joined to the roots, the fungus also enables the plant to increase its root apparatus, the absorption surface for mineral salts and water in the ground. In some types of mycorrhiza the hyphae of the fungus force their way inside the root cells to absorb the starch. The cells under attack react at a certain point and digest the fungus cells. In this way, both partners in the relationship obtain what they need: the fungus has its sugars, and the plant gains in nitrogenous substances and mineral salts.

The mycorrhizal species account for the bulk of the higher forms of fungi. It is therefore not easy to present anything like an adequate list of the species of fungus and their respective symbiotic plants.

In many instances, furthermore, a single species of fungus may associate itself with various plant species, or there may be a specific two-way relationship between fungus and plant. One of the earliest spring mushrooms in Europe, *Sepultaria sumneriana* (**416**)——a semihypogeal (subterranean) member of the Discomycetes (cup fungi)——is found only beneath cedars; *Suillus grevillei* (**251**) is a late summer and fall bolete that comes up across North America only under larches. Peat-bogs, alder woods, and sand dunes are all environments which are particularly rich in species closely associated with the plant life found in each. But there are also examples of carefully selected association with birch, larch, fir, Swiss mountain pine, eucalyptus, and rockrose. In most instances fungi are associated with the major types of broadleaf and coniferous trees. But this simplification of the overall picture often belies the environmental adaptations of fungi. The beautiful gill fungus *Hygrophorus psittacinus* (**108**) occurs in fields and meadows in continental Europe, in cork plantations in the Mediterranean region, and, in North America, in broadleaf and coniferous woods. The exquisite *Amanita caesarea* (**1**), which occurs typically in oak woods in the hotter regions of Europe, grows beneath pines in Mexico; similarly, the notorious *Amanita muscaria* (fly agaric; **2, 3**), which appears to be typical of coniferous woods, is found in stands of rockrose and eucalyptus in Mediterranean regions.

Parasitic fungi
Many plant diseases are caused by fungi, and although such diseases may seriously affect crops and harvests, and be a

major contributive factor to the shortage of food in the world, the organism responsible is rarely evident. Rusts, smuts, rots, peronospora, oidium, etc. (diseases with which farmers are all too familiar), make up a significant proportion of the fungus kingdom and cause a great deal of damage, but they are on the whole inconspicuous, and of absolutely no interest at all to the mushroom-gatherer. But some fungi do cause reactions in plant tissues and form strange and clearly visible structures called *galls*. Suffice it to mention the effect of *Taphrina deformans* on the leaves of peach trees, and the distorted shapes of maize caused by the smut *Ustilago maydis* (**420**).

Fungi are also merciless with other organisms too, including other fungi and every sort of animal, ranging from man to the tiniest vertebrate, from molluscs to insects, and worms to protozoans. Parasitic fungi are very specific: they only attack specific hosts. In some cases this specialization is extreme: the parasite of a species of beetle, for example, is only capable of growing on its elytra (wing covers) or legs. Conspicuous fruit-bodies are formed, however, by fungi which are parasitic on trees and, in some rare cases, by fungi which are parasitic on other fungi and insects.

Fungi which are parasitic on insects

There are numerous associations between insects and fungi. Of particular importance among them, because of their possible practical implications, are those which are established with the *entomopathogenic* fungi, i.e. fungi which are parasitic on insects: these fungi could well offer mankind an alternative weapon in the war against insects.

When you are out picking mushrooms you may well stumble upon entomopathic fungi. In temperate climates members of the genus *Cordyceps* are relatively common; these usually have thread- or club-shaped carpophores, often brightly colored. The commonest species, *Cordyceps militaris* (**347**) forms orange-red clubs on members of Lepidoptera and Coleoptera. *Cordyceps sinensis,* parasitic on Lepidoptera, has been used in China for medicinal purposes since earliest times.

Fungi which grow on other fungi

Some fungi grow on the carpophores of other fungi, whether still living or in the process of decomposition. In addition to the numerous molds that are parasitic on fungi, we find actual higher fungi, with carpophores, that grow on the carpophores of other fungi.

Two members of the Ascomycetes, both club-shaped, grow out of the ground but are, in effect, parasitic on two subterranean fungi: *Cordyceps ophioglossoides* (**350**) and *C. capitata* (**349**), which respectively attack the false truffles *Elapho-*

Top: Ustilago maydis *(420) growth on a corncob. Below: A cloud of spores around a fly that has been killed by* Entomophtera muscae.

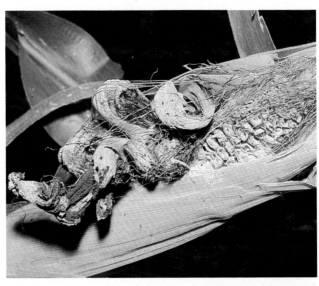

The fruit-body of Asterophora lycoperdoides
(50) growing on the cap of a Russula.

myces granulatus **(387)** and *E. muricatus.*

Boletus parasiticus **(243)**, a small yellow bolete, grows on *Scleroderma citrinum* **(376)**, a member of the Gastromycetes. The strange *Collybia racemosa* **(47)**, together with *C. cirrhata* **(42)** and *C. tuberosa*, grows on decomposing polypores and species of *Russula*. The decomposing carpophores of particular species of *Russula* and *Lactarius* frequently host small fungi with gills, *Asterophora lycoperdoides* and *A. parasitica*. And *Volvariella loveiana* grows on the malodorous *Clitocybe nebularis* **(83)**.

Some fairly common lower fungi should also be mentioned: *Sepedonium chrysospermum* is the name of a mold which attacks some boletes. The fungus produces special colored spores which cause the carpophore to be covered by a powdery yellowish parasitic mass. *Mycogone rosea* feeds and grows on the *Amanita* group, particularly *A. caesarea* **(1)** and *A. rubescens* **(11)**, which turn pink. *Hypomyces lactiflorum* causes the deformation of some white species of *Lactarius* and *Russula*, and actually improves the quality of the attacked mushroom as food. (Of course, this is by no means the case with other parasitized mushrooms.)

Saprophytic fungi
There is no organic matter which is not attacked and destroyed by fungi and bacteria. Everything that goes to form or-

16

ganic substances comes from nature and reenters the natural "economic" cycle because of the action of microorganisms. The breakdown of organic substances is achieved by fungi in the mycelial stage, i.e. as mold, even though, in many cases, this mold will produce a fruit-body.

In natural conditions the breakdown of given types of matter is carried out exclusively by specific species of saprophytic fungi. For this reason we shall now take a look at certain substrates and environments which promote the production of fungi.

Substrate specificity. Some saprophytic fungi have as their exclusive habitat no more than fragments of certain plant species. These are for the most part higher fungi of small dimension, such as *Marasimus epiphylloides,* which grows on ivy leaves, *M. buxi,* found on box leaves, *M. hudsonii,* found on the leaves of holly, *M. epidryas,* which grows on the roots of mountain avens, *Strobilurus conigena,* found on conifer cones, *Mycena seynii* on pinecones, etc. The total list would be a very lengthy one, and longer still if we started to look at the *Collybia, Marasmius, Mycena,* and *Clitocybe* groups, and the various other fungi which grow on the leaves and ligneous fragments of coniferous and broadleaf trees in general.

Even more curious and unusual organic matter plays host to the strange species of the genus *Onygena,* which grow on horses' hooves or the horns of cattle (*O. equina*), on feathers (*O. corvina*), and on the skin or fur of rodents (*O. pilifera*).

Looking for mushrooms
Equipment. Looking for mushrooms usually entails wandering over large areas of forest and grassland, and in most cases there is no guarantee that hours of walking will be rewarded by a rich crop. Clothing should be as practical as possible. The terrain in which mushrooms are found is often somewhat treacherous: it may be strewn with rocks and have dense undergrowth, and it may also be slippery and muddy. A pair of sturdy walking shoes or hiking boots is the best footwear. In addition, wear a brightly colored hat or vest or very bright-colored clothing that can be seen by game hunters.

Some kind of stick will be useful for probing the leaves covering the ground, and will also offer support in rugged terrain. You should also carry a knife so that you can pick the carpophore complete with the stipe (stalk) important for an accurate identification of the mushroom in question——and a basket in which you can arrange your mushrooms by type. The basket should be well ventilated, and strong enough to protect your crop from being bruised or damaged. The use of plastic bags, desirable for keeping green vegetables fresh, will only hasten the deterioration of mushrooms, and should *not* be used. Instead, mushrooms should be wrapped carefully in wax paper, or placed in separate open containers in the collecting basket. If too many mushrooms are placed on top of one another, those underneath will deteriorate quickly; nor should different kinds be placed together. Especially in hot weather or warm rooms, mushrooms kept enclosed can rapidly start to rot, and thus become dangerous when eaten. In fact ordinary field mushrooms or boletes that have not been carefully handled before cooking can cause stomach upsets.

What to pick. As far as the ecological functions of fungi are concerned, you should avoid picking mushrooms that are too large, or specimens which are invaded by insect larvae or *worms;* any such mushrooms would be of poor quality for eating, and, more importantly, if they are left alone they may still have some contribution to make as far as spores are concerned. If you are picking mushrooms for eating, only pick those species with which you are well acquainted. When you come upon a large group of mushrooms which you do not know, all you need do is pick one or two specimens to show to an expert. It is definitely more "respectful" to go back to pick the rest——at the risk of finding them picked by some other eager hands——than to have to throw away a basketful of inedible mushrooms which can no longer play their part in the environment.

For those who have different interests and want to look for mushrooms in order to study them, it is important to know, in advance, the sorts of species which will probably crop up in

An old tree trunk invaded by Armillariella mellea (**85**).

*A handsome basketful of field mushrooms (Agaricus arvensis, **204**).*

any given environment. The various *epigeal* fungi, i.e. those growing above ground, especially if they are fairly large, are easy to find, being at worst hidden by leaves or moss. In such cases the top of the leaf covering or litter should be raised; the same goes for mosses. Using your stick or your hands, it is easy to uncover the mushroom.

It is worth looking particularly closely at cracks in the ground, among roots and rocks, in tree stumps, among mosses, and on rotting tree trunks. All these habitats may harbor species of no concern to the gourmet but of great interest to the student. As a rule little attention is paid to species with small fruit-bodies, which are rarely included in popular books about fungi. But where these very small fungi are concerned, with the help of a hand-lens (10X magnification), a microscope, and access to the technical literature, you can often make an indentification which at first seemed impossible.

The smaller fungi grow in the litter beneath trees, on moss, and on the droppings of herbivorous animals. The dead leaves should be carefully sifted with the hands, or with a small pair of tweezers, and you will probably have to get down on your knees to do so.

You will probably often come across certain roundish fungi on this type of detailed search, and these deserve special mention because they are part of a very specific and extremely varied category: the *hypogeal* fungi, which grow below the ground.

Looking for hypogeal fungi. Looking for and gathering fungi which grow below the ground gives rise to a certain paradox between the naturalist's conscience, and the desire to gain more knowledge, because it is necessary to disturb the litter, comb the layer of humus with a knife or gardener's "hand," or even lightly hoe the ground. Any slight bulge in the ground can give away the pressure of hypogeal fungi.

A sparse covering, or absence of grass in an area beneath a broadleaf tree (oak or some other species) indicates the possible presence of the mycelium of certain hypogeal fungi. Many truffles, and first and foremost the highly prized French or Perigord truffle, *Tuber melanosporum* (**389**), in fact produce substances which impede the growth of grass.

The use of a specially trained dog or pig makes it possible to find many species of subterranean fungi, and thus avoids indiscriminate digging which damages both the root apparatus of plants and any mycelium present.

Growing sites or habitats

By growing site or habitat we mean the typical places in which fungi grow, and the most favorable places for the emergence of mushrooms are in the woods. The various types of woodland all have their own, sometimes unique, mycological stock, made up of epigeal, hypogeal, and mycorrhizal fungi, fungi which are saprophytic on leaves and pieces of wood, and parasitic fungi (even though it is not always possible to paint a decisive picture of the particular relationship between the fungus and a specific type of tree). The range of preferences may vary, and often the data which have been gathered are subsequently modified by the discovery of an hitherto unsuspected fungus-plant relationship. In this respect, and in relation to a certain habitat, the fungi can be divided into three groups: exclusive, preferential, and indifferent.

Within the various woodland formations we also find lesser environments or habitats where it is possible to find fungi that do not occur in the thickest parts of the wood or forest, but in the so-called "airy" habitats—near tracks and paths, in sunny glades and clearings, and in areas covered with bushes and other undergrowth. Deforested areas constitute a specific type of airy habitat. As well as being better "ventilated," these areas are exposed to more accentuated thermal variations and fluctuations than actual woodland, and in this respect, they resemble another important environmental category—fields, meadows, and grassland.

The term grassland describes open grassy spaces, with a fairly wide range of soils, and differing levels of moisture in the ground; they also differ in terms of altitude and the species of plants present in them. The presence in grassland areas of grazing animals introduces a very special and fairly specific habitat: dung. A large number of species, known as *coprophi-*

lous or *fimicolous,* usually small in size and shortlived, grow on dung. Here we find coprophilous species which are specialized to grow on the dung of particular animal species: *Agaricus bisporus* (**206**) grows best in horse dung, whereas *Psilocybe cubensis* occurs on cow dung.

Some selective substrates are formed by decomposing sawdust, oak- or tanbark, the sediment of apples used in cider-making, coffee dregs, and other waste matter from human activities. Burnt ground and the remains of charcoal kilns or bonfires play host to a rich *anthracophilous* flora (i.e., growing on coal), including the much-prized morels which, though not actually anthracophilous, are often found growing in burnt ground. This is one reason why peasants in various parts of Europe burn weeds and brambles along embankments.

Sand, be it in sand dunes, or by rivers or lakes, or along the seashore, where there can be fairly high mixtures of humus or any sort of chemical composition or mineralogical structure, also play host to an *arenicolous* or *psammophilous* (sand-loving) mycological flora and to other adapted species. A much more selective habitat is the upland peat bog and lowland bog or marsh which play host to species adapted to the high level of humidity and acidity in the substrate. In mountainous regions around the snow line, in ground which is not affected by thaws with the accompanying water, many members of the Discomycetes group, *Ascomycetes* (cup fingi), and other small fungi manage to survive.

The colonizing capacities of the fungi mean that they also can occur beneath the surface of the ground. Calcareous ground greatly encourages the growth of hypogeal fungi, especially the truffles, but sandy and flinty ground also plays host to interesting species.

Of the various growing habitats, the oddest of all is undoubtedly the monumental termites' nest found in Africa. The carpophores form inside these huge constructions, where the mycelium of the fungus belonging to the genus *Termitomyces* tunnels away extensively. The stipe, surmounted by a very hard, pointed cap, grows in length and energetically blazes a trail through a yard or two of compact earth. Once it has reached the outside world the tip of the cap unfolds and the cap itself, which can reach up to 60 cm in diameter, looks like a parasol placed above the sun-baked termites' nest. The termites, for their part, feed on the fungus and help to distribute new mycelium inside the nest's tunnels and galleries. The environmental conditions are so favorable to the growth of the fungus that the mycelium often tends to block the passages inside the termites' nest, with the result that the termites are forced to eject the mycelial masses which have formed. These masses, once outside, produce small carpophores, which differ somewhat from those that force their way up through the inside of the nest.

Carried around by wind and water, suspended in the atmosphere, the light spores of fungi are ready to germinate as soon as they come into contact with a suitable substrate, if the nutritional and climatic circumstances permit. So the fungi manage to live and thrive all over the world, from the surface of the land (cultivated or otherwise) and the depths of the sea to the upper strata of the atmosphere, from sand dunes along the seashore to places where the snows never melt, from human habitations to dense forests.

The nature of the ground
The ground or terrain is the result of the physical, chemical and biological alteration of the original rock.

The importance of the composition of the soil and ground for the biology of fungi has often been overestimated by various very generalized statements and assertions. It can be said, more or less categorically, that various fungi prefer certain types of soil to others, and that most fungi are affected more by the chemical nature of the soil and less by its composition. In other respects, the influence of the ground on the fructification (fruit-forming capacities) of the mycelium is evident if we consider the practices used in cultivating field mushrooms. Among the various characteristics of the soil, special importance is attached to its chemical reaction, i.e. its pH. Where the pH is "wrong" for a given species of mushroom, as it may

The life cycle of a tree, from seed to plant (left) is like that of a fungus, from spore to fruit-body (right). The various species of fungi establish mycorrhizal symbiotic relationships with green plants by attaching themselves to their root apparatus; or parasitic relationships, where they actually penetrate the plant through cracks, etc.; or saprophytic relationships, where they transform wood and dead leaves into mineral substances.

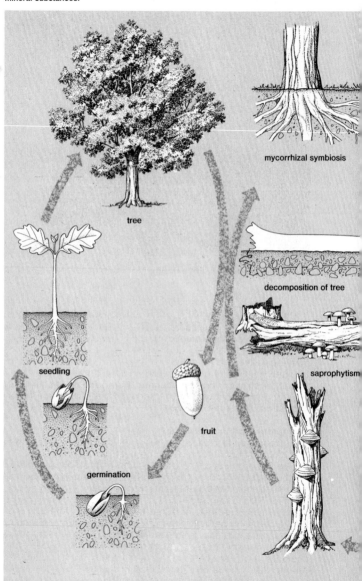

mycorrhizal symbiosis

tree

decomposition of tree

saprophytism

seedling

fruit

germination

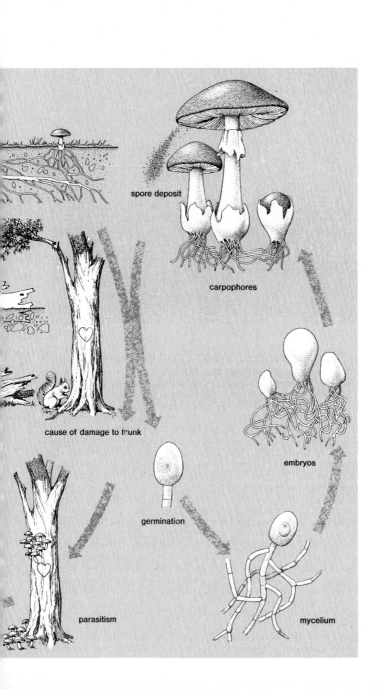

spore deposit

carpophores

cause of damage to trunk

embryos

germination

parasitism

mycelium

An unusual crowd of field mushrooms (in the foreground,
Coprinus comatus, **224**).

be in a park setting, the mushroom may not occur despite the presence of its mycorrhizal tree host.

As far as the pH is concerned, ground can be neutral, basic, or acidic. Correspondingly, the various species of fungi may be considered either acidophilous or basophilous even if, in reality, the optimum pH of many fungi is close to a neutral level. In some cases trees can modify the pH of the ground in which they are growing. This applies for example, to the chestnut tree, a typically acidophilous species: if planted in calcareous ground, it tends to acidify the surface soil. In fact, plants can generate particular environmental factors such as the accumulation of humus (a compound of organic substances mainly of vegetable origin), and they can produce *rhizospherical* phenomena, i.e. phenomena connected with the volume of earth modified and affected by the root apparatus. These can create actual islands in the midst of the surrounding environment. Many fungi become adapted to such changes; others do not. The French or Perigord truffle (*Tuber melanosporum,* **389**) grows in markedly basic, calcareous soils, whereas the white truffle (*Tuber magnatum,* **391**) prefers clayey, marly, acid soil. On their respective preferential grounds these two truffles can become associated with numerous species of trees, some of which are symbiotic with both the French and the white truffle. But the symbiosis with one or other type is established by the type of soil. In truffle-growing areas, what is more, the French truffle does not extend into the denser areas of the wooded

Not even furniture is safe from invasion by lignicolous fungi such as Serpula lacrimans.

formations made up of its symbionts because the accumulation of dead leaves, which creates humus, acidifies the upper strata of the soil.

When to look for mushrooms

It is well known that periods of dry weather or intense cold are not favorable for the development of the fruit-bodies of fungi. But this does not mean that no fungi are in the process of reproduction. Many sorts of hypogeal fungi, truffles and other types continue to grow and ripen during winter and summer, and many species living in very humid environments find the conditions suitable for their development during the hottest and driest periods of the year. The best conditions for the occurrence of a massive production of carpophores have to do with the presence of humidity in the nutritional substrate and in the air, and with a mild temperature.

It is difficult to put precise values to these conditions, not least because there exists, for each species, an optimum temperature and an optimum humidity, just as there exist upper and lower temperature and humidity limits outside of which the fungus cannot form fruit (sometimes it cannot even survive). The climatic pattern of the various regions thus affects the period when fungi appear as well as their geographical distribution. On the whole, in temperate climates, the highest production of carpophores occurs in late summer and early autumn, when the atmospheric precipitation lowers the August temperature and raises the humidity level of the ground. These conditions usually occur in successive stages, moving from the mountains to the sea, and from north to south.

In the Mediterranean regions there are two periods of production which are sometimes very similar as far as the fungus composition is concerned: one occurs in the spring, the other in the autumn. In the case of the autumnal period, rain helps to lower the summer heat; in the spring, the sun heats up the cold, wet ground. And sometimes, if the spring temperature and humidity levels are optimal, autumnal species will emerge along with the typical species found in the spring. In desert areas or in sand dunes by the sea, on the other hand, we find species whose appearance is associated with the presence of minimal amounts of water. Here, too much water in the ground may impede the germinative capacities of the spores or the potential fruit-forming capacities of the fungus. Thus in excessively rainy seasons many of the species normally present will not form fruit.

During very dry seasons the most abundant species belong to the group of saprophytic fungi, especially those saprophytic on wood, since this substrate retains the moisture necessary for a longer period than ground leaves and other surface substrates.

A dry period coming after a particularly wet one will encourage the fructification of those mycelia which have been able to develop properly. There is plenty of information and evidence to confirm this. Also, it has been maintained that in hot or warm regions in particular, the fruit-forming capacities of fungi are due not only to rain but also to the dry period preceding the rainy one. The morphogenetic action of dry weather has led to the formulation of what is known as the "theory of fructification by endurance." This theory finds supporting evidence in the ways in which the champignon and other cultivated mushrooms are grown.

Temperature is the other basic factor of any climate, although for most fungi the temperature of the substrate is more important than the air temperature. Thus sunshine and the sun's heat play an important part in the fructification of fungi, and here too the requirements of the various species are fairly diverse. Alongside species which prefer sunny, open spaces, we find others which seem to shy away from the direct rays and heat of the sun. But almost all the fungi disappear as the cold season approaches. Apart from the hypogeal species, which find protection below the ground, only a few other fungi continue to grow during the cold winter months; these include a few *lignicolous* species (growing on wood), among them the well-known oyster fungus, *Pleurotus ostreatus* (**113**), and *Flammulina velutipes* (**46**), various Gastromycetes such as the *Tulostoma* and *Cyathus* groups, and certain minute species which grow in mosses.

In addition to humidity and temperature we should also

briefly mention light and wind. There are many species of fungi which can form fruit in the complete darkness of mines and caves. For most species, however, a minimal ration of light is needed both to form carpophores and for the normal development of the fruit-bodies. In fact some of the species grown in the dark do not always grow properly. Other species fail to form fruit if they do not have adequate light. For some species strong bright light can be harmful, whereas high-altitude and desert fungi thrive on it.

The influence of the wind is well known to all mushroom-gatherers. In fact a windy spell——even if it only lasts for one day——can irreparably harm the fruit-forming capacities of fungi. The ventilation causes the water present in the ground to evaporate quickly, and the mycelium dries out, as do the embryonic forms of the carpophores. The cuticle of already developed fungi wrinkles and splits. All the good work done by the rains is suddenly undone.

The life of fungi and the fairy-ring
The mycelium which gives rise to the carpophores has a fairly short-lived existence. The mycelium of coprophilous species lives only a few days and the subsequent production of carpophores is due to the formation of new mycelia orginating from the spores dropped by earlier fruit-bodies. In the case of the short-lived species, such as the coprophilous fungi, the life cycle is somewhat speeded-up and in some cases the carpophore itself survives no more than a single day. From the moment when the saprophytic species make contact with the substrate and germinate, their life span is closely correlated with the nature of the substrate. The *xylophagous* (feeding on wood) species live until the cellulose of the host structures has completely decomposed and has been removed by other rival organisms. Among the species which grow on wood, those with the longest life span belong to the family *Polyporaceae* (polypores) which start to attack trees that are still living. The sometimes exclusive presence of a *polypore* causes the wood to rot slowly, eventually killing the tree; when the tree dies, this also marks the fairly swift demise of the parasite as well, it being increasingly restricted in its nutritional needs by other new xylophagous species. The mycorrhizal species seem to have an even longer life span, and they are also more prudent, as it were, from a survivalist point of view. The symbiotic relationship with plants, though subject to constant competition and an almost "dialectical" pattern, can mean a life span of several decades.

Sometimes you will come upon long strips of ground or circular areas where the grass is greener and more lush. In these areas, at the right time, numerous carpophores will grow, also arranged in a line or circle: the "fairy-ring." These fairy-rings are caused by a natural phenomenon which, sadly, has nothing to do with magic: the responsible party is merely a specific and tiny fungus spore. The hyphae produced by germination spread in all directions (although the extent depends on the makeup of the ground) in their attempt to colonize and conquer more and more new territory with available nutritional substances. The hyphae grow at the tip, ramifying continually with the result that as the colony fans out at the edges, it grows old and dies in the center. In this way a circular strip is formed, extending outward each year. If some obstacle impedes the circular formation, the hyphae advance on just one front, first in the shape of an arc and later, as the seasons progress, in a fairly fixed line. The mycelium in the ground carries out a function similar to that of the bulb and stolons (runners) which are primed to produce a flowering plant each year. In fact the mycelium of these formations represents an enduring and subterranean form of the fungus, sometimes dormant and sometimes active, which, together with the spores, ensures the survival of the species. The growing end of the mycelium eliminates the waste matter, thus fertilizing the edge of the circle——and this is where the grass grows so luxuriantly. The grass in the inner part of the circle is thin and stunted, either because it is poisoned by the substances deriving from the decomposition of the dead hyphae, or possibly because the old mycelium produces antibiotic substances.

The life span of fairy-rings is quite long. The mycelium is only killed off by environmental changes, such as in a valley invaded by the overflow from a hydroelectric dam, or on a hillside or mountain undergoing reforestation. In some areas of the western United States, where man's interference has been very limited hitherto, enormous fairy-rings have been found; the age of some of them has been calculated in hundreds of years. The species which most often give rise to these formations belong to the genera *Agaricus, Amanita,* and *Lepiota,* and the best-known fairy-ring mushroom is *Marasmius oreades* (**31**). Fairy-rings can also form in woods, with other genera responsible for them, but since there is no grass they are only visible when the carpophores appear.

FUNGI AND MAN

The presence of fungi has been known for thousands of years. The first hints of our mycological knowledge, relating to plant diseases, are recorded as far back as 1200 B.C. in the ancient Vedas; and the effects of poisonous fungi were dealt with in an epigram written by Euripides in about 450 B.C. The larger types of fungi (macrofungi) have been of interest to mankind from very earliest times; the use of their prodigious metabolic capacities got under way when people started to develop a taste for wine or leavened bread. Even today, in a world undoubtedly more aware of things scientific, few people realize how their lives are closely bound up with the presence or activities of these organisms; and few people realize how not a day passes, in a manner of speaking, when each one of us is not

helped or harmed, directly or indirectly, by these "citizens" of the microcosm surrounding us.

Fungi play an extremely important part in the slow but continual changes taking place in both nature and society. They are the agents responsible for the breaking down of much of the organic matter (saprophytism), and as such they cause huge amounts of damage by destroying foodstuff, hides and skins, fabrics, wood, and other consumer goods, as well as books and works of art. In addition, the fungi are the cause of the majority of known plant diseases, as well as many diseases which affect animals and man (parasitism). On a positive note, fungi are at the root of many industrial processes, which include some of man's most ancient activities, e.g. the production of wine, beer, and bread, and the distillation of alcohol. And their use has been gradually extended to the commercial production of antibiotics, vitamins, organic acids, enzymes and other alimentary and pharmaceutical products.

In our agricultural activities, too, fungi are both a help and a hindrance. Though they are responsible for considerable damage to crops and harvests because of the plant diseases they cause, they also make agricultural land more fertile by means of the degrading action they have on organic substances. This action is also called "mineralizing," and because of it nutri-

A typical example of a "fairy-ring," a completely natural phenomenon which used to be a source of much superstition.

tional elements are formed which can be reused by certain green plants.

The cultivation of mushrooms

The observation that certain edible mushrooms grow naturally on certain decomposing organic matter and the desire to produce a tasty source of food have led to the development of various techniques for the cultivation of mushrooms, and in some cases they have been fairly profitable. But the cultivated species have never been the most prized by gourmets. Their favorite types are still the boletes, specific species of *Amanita* and truffles; all these are mycorrhizal fungi, and their carpophores cannot be developed without the right symbiotic plant.

To date the only successes have been with the saprophytic species, i.e. those which feed on dead organic matter. Among these the best-known is undoubtedly the champignon, *Agaricus bisporus* (**206**), which has given rise to a very thriving agricultural industry. Each year millions of pounds of these mushrooms are produced for direct consumption, for the canning industry and for the preparation of soups and sauces. This activity dates back to the 17th century in Paris, during the reign of Louis XIV: to begin with the mushrooms were cultivated in outdoor gardens; subsequently, with the use of underground tunnels and stone cellars, it turned into a fully-fledged industry. In the 19th century, in fact, 1,500 miles of tunnels and underground chambers were used for the cultivation of champignons. The soil used was made up of horse dung, a substrate on which this mushroom grows naturally.

In recent decades, following the huge development of the microbiological industry, the process has been enormously improved and the production of mushrooms in a controlled environment has become an expression of highly advanced technology. Today horse dung still constitutes the basis of the culture soil. With a given quantity of straw and droppings, the dung initially undergoes a process of natural, though controlled, fermentation, first out of doors and then in an enclosed area where it is enriched with nitrogenous sugar substances and vitamins required for the development of the field mushroom. After variable periods of time, depending on the composition of the substrate, now called "compost," sowing takes place, using cereal seeds covered with *Agaricus bisporus* mold. Today two main varieties of field mushrooms are used: white and brown. The first is better suited to canning, while the second, with its hazel-colored cap and small brown scales, is more widely used in Europe for direct consumption, because the flesh is firmer and tastier. The American preference for the white variety is an expression of the same prejudice reflected in the choice of white eggs, bread, and sugar.

The spores are mixed with the compost and left in very humid chambers at a temperature of 75–77° F (24–25° C). After a couple of weeks, when the mycelium has spread throughout the compost, the soil——arranged variously in long mounds, on platforms, or in boxes piled on top of one another——is covered a few inches deep with a mixture of crushed stone and peat, and sterilized with a solution of formalin. The temperature is lowered by about 16° F (8–10° C) and after ten days or so the first fruit-bodies are produced; these will reform each week, producing "flushes." After five or six flushes the culture is renewed.

The cultivation of the oyster mushroom, *Pleurotus ostreatus* (**113**), and closely related species is also highly developed, especially in the Orient, and they are found in western markets in various varieties. *Volvariella volvacea* (padi-straw, **159**) and *Lentinus edodes* (shii-take), common in Oriental cuisine, can also now be obtained in specialty shops in North America. For other species, such as the morels, production techniques still have to be improved, while, with the constant demand that currently exists, production lines are being launched for *Stropharia rugosoannulata* (**210**) and *Agrocybe aegerita* (**195**).

All the saprophytic species, especially those growing on wood, can easily be cultivated even though in many instances they do not command a large market because of their meager amount of flesh. But they can be cultivated on a small scale or at home by infecting unhealthy plants or dead stumps with

Left: A giant fruit-body of Polyporus squamosus *(309). Right: An appetizing plateful of grilled ceps* (Boletus edulis, *240).*

fragments of the cap. The species which lend themselves best of all to this practice are *Armillariella mellea* **(85)**, *Flammulina velutipes* **(46)**, and *Pleurotus ostreatus* **(113)**.

The cultivation in mycorrhizal fungi which need to live with a host plant in order to produce their fruit-bodies, is a demanding undertaking. This is known as "indirect cultivation." The only mycorrhizal symbiotic type which is cultivated quite extensively in France—and now in Italy——is the legendary French or Perigord truffle, the high quality and price of which amply justify the cost of installing artificial truffle beds and the long years of waiting before the carpophores can be picked.

In centers of modern truffle cultivation production is followed extremely closely and nothing is left to chance. The acorns are sterilized on the outside in a chemical bath or dip, and are then put in direct contact exclusively with truffle spores. The seedlings which develop in small bags filled with sterilized earth will have the truffle as the only symbiotic organism with which to associate. Once planted out in soil with the right characteristics, and with due care, they will start to produce their first truffles after about ten years.

FUNGI AND ANIMALS
The cultivation of fungi is not an exclusively human activity; in fact various insects need fungi for their survival and have thus

become exceptionally efficient mushroom-farmers.

A good example of the relationship between fungi and animals is given by certain ants in tropical America of the genus *Atta*. These ants take green leaves inside their nest. In special rooms the leaves are finely shredded and neatly stacked. Mycelium then develops on this organic substratum and makes up the food of the ants. It is also probable that African termites have a very similar relationship with fungi of the genus *Termitomyces* which grow in termites' nests. In addition, one constantly comes upon fungi that have been to some extent eaten by animals, and despite what is commonly believed, not all of these fungi are edible species. Slugs, for example, casually devour the fearsome death cap, *Amanita phalloides* (**4**); in fact their tolerance to the *Amanita* toxins is about 1000 times higher than man's. In the animal kingdom the most impressive consumers of fungi are the insects, many of which spend their larval stage as worms inside fungi. How many times is one reluctantly forced to leave plump boletes, snow-white field mushrooms, and priceless truffles to the voracious appetite of larvae!

As well as slugs and insects, some mammals also seek out and feed on fungi. Pigs and wild boar are particularly fond of fungi, as well as a large number of other herbivorous animals, both wild and domestic, who vary their diet by eating fungi.

Deer are so voracious for fungi that at times they become our competitors for the available supply.

THE TOXICITY OF FUNGI

There are a number of false beliefs about the toxicity of fungi: that a poisonous one will cause silver to tarnish and garlic to blacken; that if a cap has been eaten by a slug or animal it can safely be eaten by a human; that any dried fungus is safe, because toxins lose their strength as the carpophore dries; that no lignicolous (i.e., growing on wood) fungus is deadly; and that symptoms of poisoning will appear immediately upon ingestion. The facts are otherwise. Although many fungi will cause a stomach upset within 30 to 60 minutes, the deadly *Amanita phalloides* (**4**), *A. verna* (**7**), and *A. virosa* (**8**) do not produce noticeable symptoms until 8 to 12 hours after ingestion. *Galerina autumnalis* (see **note** at **196**) which contains the same toxins as the deadly species of Amanita (amatoxins), is common on wood. Although certain fungi do lose their toxicity upon drying, other fungi, especially those containing amatoxins, are poisonous no matter how they are processed. No reliable conclusions about the edibility of a fungus can be drawn from seeing that it has been eaten by a slug, insect, or any animal. And old wives' tales regarding the nature of poisonous fungi are nonsense.

The toxicity of fungi is a genetic characteristic of the species, and the only way of establishing the edibility of a species is to have a thorough knowledge of its morphological features.

In addition, the dispersal of large amounts of highly toxic substances like pesticides (fungucides and insecticides), can make any fungus which may come into contact with them dangerous. Similarly, fungi which grow in areas where there is heavy traffic or a high incidence of industrial pollution may also become poisonous, even though they are normally excellent to eat. Here the fungus acts like a sponge and concentrates in its carpophore certain highly dangerous metals such as lead and mercury. These pollutants accumulate in the soil and are absorbed and concentrated by the mycelium of the fungi.

Cases of poisoning by fungi are mainly due to substances produced by and contained in the fruit-body; these substances may act immediately at the gastrointestinal level, causing violent stomach upset which in turn causes the intestine to evacuate its contents, thus preventing a complete absorption of the poison; or they may enter the bloodstream as toxins and affect various organs, in some cases fatally, or irremediably. In the first instance the outcome is almost always positive, as long as there are no complications, but in the latter, irreparable damage to vital organs (liver and kidneys) can occur, and there is some possibility of death. In all cases requiring hospitalization, whatever specific remedies are used, fluids, sugars, and salts must be replaced and maintained.

Lignicolous fungi are easy to cultivate. This photo shows the fructification of Agrocybe aegerita *in a laboratory.*

Many of the organic compounds responsible for the toxicity of fungi are known to us because they have been isolated and identified chemically; and for some we have satisfactory knowledge about the way they act. These studies have made it possible to elaborate remedial methods which have shown themselves to be fairly effective in cases of mushroom poisoning; there have even been successes in cases of poisoning by the deadly death cap, *Amanita phalloides* (**4**) and other similar species, and good hospital care and fluid maintenance have helped to reduce the fatality rate caused by this kind of poisoning.

THE EDIBILITY OF FUNGI

Not all the fungi in this book are known to occur in North America, and those that do represent only a small portion of the 10,000 species that occur here. Many of the fungi in this book, however, are among our most common species. Except for a few distinctive kinds fungi are difficult to identify. Because a few common species are deadly and many are known or believed to cause various degrees of poisoning, it is necessary to exercise the utmost caution in determining the identification of a collection.

Be sure to collect the whole fungus: don't leave a portion of the stipe in the ground because one of the identifying characteristics of a deadly *Amanita* is the cup about the base of the stipe.

Keep all species in separate containers to avoid contamination.

Make a spore print before attempting to identify gilled fungi: often the color of the gills gives no indication of the color of the spores.

Keep fungi refrigerated until eaten, and eat only those fungi you know to be edible and that are in prime condition.

Never eat wild fungi raw.

Always observe moderation in eating fungi. Eat a very small amount the first few times you try a new species to be certain you are not allergic to an otherwise edible fungus.

The gilled fungi, while very common during wet periods throughout the season, are the most difficult to identify safely for eating. The beginner should be content to attempt their identification, but should reserve for the dinner table those more easily identified non-gilled fungi. These include boletes (239–258), except for those with red pore mouths and those that are bitter; chanterelles (154) and black trumpets (156); the cauliflower coral (329); morels (282–288); fleshy polypores (325–327); puffballs (368–373), and tooth fungi, especially the bear's head (314).

Top: The cultivation, on straw, of Pleurotus ostreatus *(oyster fungus, 113) out of doors. Below: Harvesting cultivated mushrooms or champignons (*Agaricus bisporus, **206**) *in a modern, enclosed environment.*

Beware: Fungi that have been eaten by insects and slugs may be poisonous! (But not because they were eaten by insects and slugs.)

Once one acquires proficiency in using the keys and reading descriptions, one may want to taste some of the choice gilled fungi. Some of the more easily identifiable of these include: the almond-or anise-scented species of *Agaricus* (202, 204); the honey mushroom (85); *Lactarius volemus* (127); the oyster (113, 114); the parasol (20); and the shaggy mane (224).

HOW TO STUDY FUNGI

Attempts to identify a fungus by looking for a picture of it in a book are not usually very successful. This is first and foremost because no book exists which can boast that it includes or illustrates all the fungi; it is also because not all the fungi have been drawn yet, especially in color, let alone photographed. This situation is sometimes aggravated by the poor accuracy of some illustrations and reproductions, or by the fact that the specimens photographed were not in a perfect state, or because the printing process has altered some of the colors. In addition, one should keep in mind that while certain mushrooms are relatively easy to identify, most are not; and there are a great number of look-alikes. In order to avoid any unpleasant experiences and in order to have the beginnings of a valid identification method, you must inspect your fungi with a critical eye. In particular, the structures which form the carpophore must be closely examined. Comparison of the specimen with a colored illustration can help to *substantiate* a hypothetical identification.

It is also important to take note of the environment in which the fungus is picked. As well as trying to fit the fungus into the systematic mycological classification, it must also be fitted into its natural surroundings; in other words we must understand its ecological role. For this reason it is important to know whether the fungus grows directly out of the ground, whether it has grown on a pinecone or on a decomposing branch, or on a tree trunk or on the remains of another fungus. In addition one must not overlook the fact that the fungus may have been picked in a forest or in grassland, and note must be taken of which trees, grasses, or mosses were present in the area of the find. It is also worth noting the type of ground in or on which the fungus was growing. There are many factors to be taken into account, and in some cases they can be of fundamental importance for a correct identification of the species of fungus which has produced the carpophore.

The identification of a fungus, i.e. the species to which it belongs, will follow this sort of logical sequence:

1) Observation of the morphological features of the carpophore (shape, possible remains of veils, form of the hymenium, color, cuticle, consistency of the flesh and the possible production of latex at the point where it has been broken off, and habitat). This first series of observations should indicate the genus to which the fungus belongs, or, at the very least, to certain similar genera which will only be able to be ascertained on the basis of their microscopic features.

2) Color of the carpophore, presence of typical veil remains, shape of cap and stipe, color of flesh, any possible color change, color of latex (where applicable), smell, taste, and habitat form the second level of observation. Once these data are known, the systematic position of the fungus can be further isolated, and in many cases it will be possible to identify it on this basis.

3) With macro- and microchemical reactions, and observation by microscope you will be in possession of virtually all the factors needed to name the precise fungus in question. By comparing your data with a detailed description and a color chart your identification should now be complete.

The ordinary individual, unfortunately, does not always have the right technical means at his or her disposal for microscopic and chemical examination of the carpophore, nor the right bibliographic references (which are scattered through dozens of scientific journals, or incompletely compiled in books and publications often written in foreign languages, usually impossible to find in bookshops, and too expensive). We have included a bibliography which we hope you will find useful, and suggest that you contact university botany departments, museums of natural history, and botanical gardens. The North

THE TOXICITY OF FUNGI

Syndrome	Species	Incubation
Gastro-intestinal	*Agaricus xanthodermus* (203) *Boletus satanus* (245) *Chlorophyllum molybdites* (see 21) *Entoloma sinuatum* (163) *Hebeloma crustuliniforme* (187) *Lactarius torminosus* (131) *Naematoloma fasciculare* (214) *Omphalotus olearius* (86) *Pholiota squarrosa* (191) *Ramaria formosa* (330) *Scleroderma citrinum* (376) *Tricholoma pardinum* (65)	30 min.–2 hours (up to 3 to 6 hours)
Botulinic	Altered carpophores	1–4 hours
Amanitin poisoning	*Amanita phalloides* (4) *Amanita verna* (7) *Amanita virosa* (8) *Galerina autumnalis* (see 196)	8–24 hours (up to 48 hours)
Amanitin-like poisoning	*Lepiota helveola* (see 25) and related small species	5–15 hours
Orellanine	*Cortinarius gentilis* (see 178) *Cortinarius orellanus* (186) *Cortinarius speciosissimus* (178)	3–14 *days*
Gyromitra (false morel poisoning)	*Gyromitra esculenta* (279) and possibly related species	6–12 hours, sometimes after 2 hours
Psilocybin (psychoactive)	*Gymnopilus spectabilis* (198) *Psilocybe cubensis* (see 211) and related species some species of *Panaeolus* and *Conocybe*	30–60 minutes
Muscarine poisoning	*Clitocybe dealbata* (see 31) *Clitocybe phyllophila* (78) most species of *Inocybe* (168–171)	15 min.–4 hours
Muscimol (fly-agaric poisoning)	*Amanita muscaria* (2, 3) *Amanita pantherina* (10)	30 min.–2 hours
Coprine	*Clitocybe clavipes* (81) *Coprinus atramentarius* (223)	30 min. after drinking alcohol, up to 2 days after eating mushroom
Immune Injury	*Paxillus involutus* (236)	gradually acquired hypersensitivity

Principal Effects	Duration
affects the gastrointestinal apparatus (vomiting, diarrhea, colic pains, cramps)	6 hours to 2 days, cases of poisoning by *Chlorophyllum molybdites*, *Entoloma sinuatum*, *Omphalotus olearius*, and *Tricholoma pardinum* may be more severe, and may require hospitalization
affects the gastro-intestinal apparatus and nervous system	3–5 days, sometimes fatal
severe vomiting, diarrhea, and cramps for 12–24 hours, followed by a brief remission and the onset of kidney and/or liver dysfunction or failure	10–20 days; when fatal, death coming usually 4 to 7 days after first symptoms
intestinal symptoms, muscular cramps, sweating, congestion of internal organs	not properly known
damage to liver, kidneys, spinal cord, with intestinal upset, fever, chill, headache, muscular pain	1–5 *months*, often fatal
bloating, abdominal pains, vomiting, jaundice, bloody diarrhea	5–10 days, death can occur after 2 days; dose related
mood change; pleasant or apprehensive; unmotivated laughter, visions	4–6 hours, sometimes longer
dizziness, blurred vision, shivering, profuse sweating, salivating, nausea, circulatory disorders	12–24 hours, sometimes 2–4 days; may require hospitalization; death rare
dizziness, hyperactivity, deep sleep, delirium, sometimes sweating	1–2 days
flushing of upper body, tingling in extremities, hypotension, nausea	2–4 hours, rarely serious
diarrhea, vomiting, cardio-vascular upsets, allergic reactions (anaphylactic shock), massive hemolysis	2–4 days, rarely fatal

American Mycological Association (NAMA), a national association of people interested in collecting, studying, photograph-
ing, and eating mushrooms, will provide you with the ad-
dresses of the local mushroom clubs in your area. Write: The
North American Mycological Association, c/o The New York
Botanical Garden, Bronx, New York 10458.

The Classification of fungi

Because of the fairly general presence of a wall which gives
the cells a rigid structure, the fungi have been considered
plants without chlorophyll and with an invisible method of reproduction—members of the plant kingdom. But there are
many mycological systematists who tend to put the fungi in a
completely separate kingdom, with the same status as the
plant and animal kingdoms. A fairly wide variety of organisms
are considered members of the fungus family.

This book only deals with fungi which form fruit visible to the
naked eye, the *macromycetes* (large fungi). Without going into
too much detail we can say that the fungi which produce carpophores fall into two major groups: the Ascomycetes and the
Basidiomycetes. In the first class the spores are produced inside microscopic sausage-shaped organs called *asci;* in the
second the spores are outside the microscopic club-shaped
reproductive organ, the *basidium.*

The number of species.   It is hard to say how many fungi
there are. The lack of information and communication, particularly on an intercontinental scale, has meant that the same fungus sometimes has several names, and that different fungi
sometimes have the same name. Despite the impossibility of
giving a precise figure, there can be no doubt that the number
of fungus species runs into tens of thousands.

The aim of this book is to give some idea of the great com-
plexity of just a small area of mycology, the area occupied by
the macrofungi (large fungi), and the vast variety of forms
which can be found among the members of this group. For
these fungi the essential criterion for their classification lies in
the fruit-body, an ephemeral and fairly changeable structure.
The large fungi found in any country with a temperate climate
total many thousands of species. Becoming an expert on the
macrofungi in your particular locale will be a challenging and
rewarding lifetime pursuit.

The geographical distribution of fungi.   The spores of
fungi are scattered all over the surface of our planet. Given the
right environmental conditions they can germinate and form a
fairly enduring mycelium; if these conditions combine with particular climatic conditions, fruit-bodies can be formed; these
are the only elements which enable us to identify fungi. In the
brief period of time in which the carpophores are visible, they
must cross paths with a keen and well-versed mycologist in

The structure of a fungus with cap and stipe.

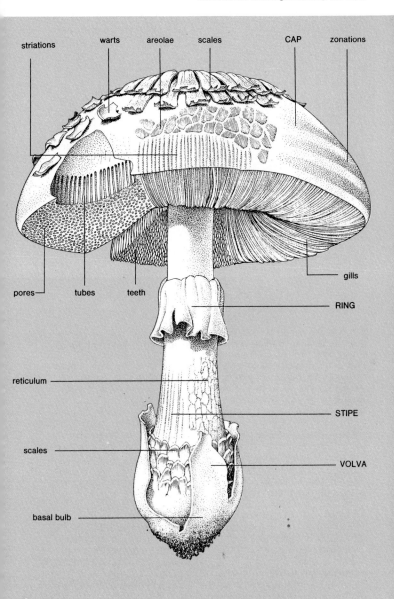

striations warts areolae scales CAP zonations

pores tubes teeth gills

RING

reticulum STIPE

scales VOLVA

basal bulb

Model of a spheroidal fungus.

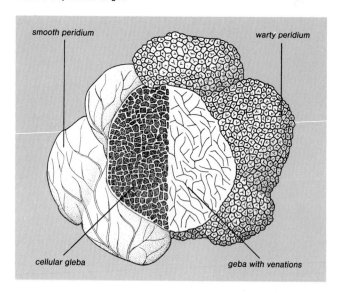

smooth peridium

warty peridium

cellular gleba

geba with venations

order to be recognized and included in the mycological flora of a country. The likelihood of all these factors happening in the order described is understandably extremely slim, so the absence of a given species in a continent or region can almost certainly be put down not to an actual fact, but to ignorance. All it takes is an area replanted with exotic species of trees, imported merely as seeds, to completely alter the flora by awakening, in the soil, the hordes of spores carried on air currents from the country of origin of the trees introduced. Australian eucalyptus species, for example, used for reforestation in the hot temperate zones of both hemispheres, produce an Australian mycological flora even when planted in Italy or California. In any event, with a long inventory to help us, we can say that the majority (70–80%) of fungus species occurs in the same habitats more or less all over the world.

The color of the flesh

The color of the flesh of fungi is an important feature. It is examined by cutting the carpophore in two, from top to bottom. The inside of the carpophore, the "flesh," can be of various colors, and in some instances will change color as the fungus matures, as in the puffballs, which begin as white and apparently undifferentiated but which color dramatically as they mature.

Top: Cross section of a gill fungus seen through a microscope; the darker elliptical structures are the spores. Below: Cross section of the gleba of a puffball under the microscope; the spores are the small spheres which form the darker stratum.

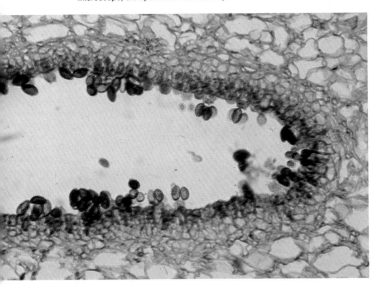

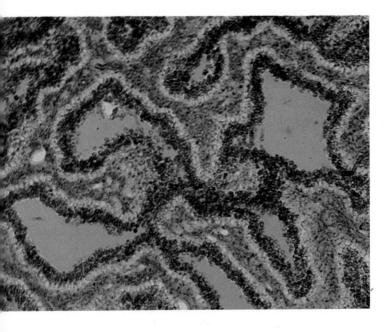

A Coprinus *basidium with two mature spores (on left),
seen under the microscope.*

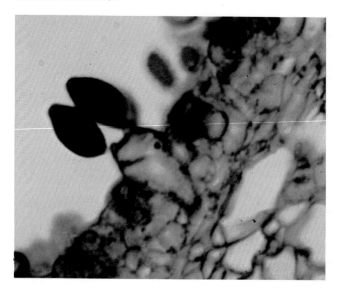

As a rule the color of the flesh is important for the identification of cap-and-stipe fungi (Agaricaceae, Boletaceae, Hydnaceae) and slightly less so for the Polyporaceae and Thelephoraceae. Most fungi have a fairly white flesh which stays white when cut, in other words it is unchanging, or does not react. But special attention should be paid to specific parts, especially beneath the cuticle and on the sides of the stipe and at the base, which may have typical, though often only slight colorations. In many cases, however, the color of the flesh changes when the fungus is touched or broken. The color change often also affects the hymenium. There are also many fungi which have a flesh which is often the same color as the whole fungus, such as all-yellow (*Cantharellus cibarius* (**154**).

The color of the latex or milk
Some fungi produce large amounts of juice, called *latex* or *milk*, when even lightly scratched or damaged. This is a major characteristic of an important genus of fungi called *Lactarius*. But it also occurs in a few species of the genus *Mycena*, a large group but one of little interest to the mushroom-gatherer because most of the members are so small. In the *Lactarius* group the milk may be a constant or variable white, or strikingly colored from the start.

The spores of the Ascomycetes are contained inside asci, which are sausage-shaped organs.

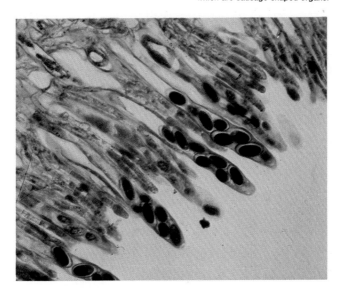

Consistency

The consistency is the degree of resistance, compactness, and firmness which the flesh of a carpophore demonstrates when broken or cut. There are four major categories, which can in turn be divided into a fairly wide range of subcategories. The following is a suggested list for the various degrees of consistency, with the major categories in italics: viscous-mucous-*gelatinous*-spongy-soft-*fleshy*-fibrous-compact-*friable* (crumbly)-waxy-caseous (cheesy)-chalky-sticky-elastic-*coriaceous* (leathery)-suberose (corklike)-ligneous (woody).

Taste and smell

In many cases the taste of the flesh enables one to find a path through the maze of fungus species, and sometimes to identify a species straightaway. A small piece of the cap should be nibbled until the taste emerges, and then spat out. It is not recommended that you nibble any species of *Amanita* or any deadly mushroom; in these cases the information you acquire by tasting the mushrooms in no way aids in their identification, which can be ascertained by examining the color, shape, and surface features.

Most of the fleshy fungi have a rather insignificant taste, but some are bitter or pungent, or have a specific flavor, like the peppermint flavor of *Russula lepida* (**139**); others may be

*Cross sections of carpophores. Left: a puffball (Vascellum pratense, **374**), showing the fertile part (the gleba) and the sterile part (the subgleba). Right: a morel (Morchella conica,**285**).*

strikingly sweet, because of the presence of mannite or manna sugar, like *Clavariadelphus truncatus* (**339**). In the *Russula* group the sharp or sweet taste is a very important factor in identifying the species. In the similar genus *Lactarius* the flavor of the milk is as important as that of the flesh itself.

Another very important organoleptic characteristic is odor. When ''smelling'' your fungus, place it in the slightly cupped palm of your hand, then hold it close to your nostrils and smell it all over, taking short breaths at brief intervals. Many genera of fungi have such distinctive odors that it is sometimes possible to identify a species by that one characteristic.

Spores
The spores are the fungus's means of propagation and survival. The quantity of spores produced by each carpophore is incredibly large. A typical field mushroom, for example, produces about 16 billion spores, released at a rate of about 100 million an hour. A large puffball with a diameter of around 12 inches can produce up to 700 billion spores. The minute spores of the puffballs are scattered by the wind, and, given their infinitesimally small weight, it is quite possible for a spore to fly around the world on air currents before finally coming to rest. Spores are an important factor in the study of fungi, and the systematics of fungi and their classification are based on spore characteristics. ·

In order to study the method of formation, the size, the morphology, and the chemical reactions of spores you will need a microscope and considerable specialist experience, two things out of reach of most of us. Apprentice mycologists will find in the spores, or more correctly in the spore-print, a fairly important element on which to base their verdicts; and the spore-print will help them find their way among the thousands of species of fungi with caps. The spore-print is a deposit of spores in mass. It is obtained by placing a cap with the hymenium facing down on a sheet of white paper or a piece of glass. After a few hours (sometimes not until the following day) a layer of spores will be deposited. This way you can determine the real color of the spores, which will fall into four or five broad types: white, pink, ochre, and blackish-violet; often this last grouping is divided into two groups, purple and black. In the large corresponding groups——the leucospores (with white to pale yellow spores), the rhodospores (with pinkish spores), the ochrospores (with ochre to brown or rust-brown spores) and the iantino- or melanospores (with violet or black spores)——there are very few exceptions. The most notable exception is the poisonous green-spored *Chlorophyllum molybdites.*

The color of the spores can be somewhat roughly detected by looking at the color of the gills of a mature specimen, but sometimes the color of the gills may in fact hide that of the spores. A violet fungus such as *Clitocybe nuda,* for example, has pale pinkish buff spores which do not show up against the surface of the bright lilac-violet gills.

Macrochemical reagents
Various chemical substances produce specific color reactions on contact with the various structures of the carpophore. (See Bibliography for references.)

The use of reagents, which are *not* used to distinguish between edible and poisonous fungi, has opened up a new frontier in mycological systematics. The discovery and formulation of new reagents, is a very slow process and they are not infrequently stumbled upon by chance.

The mycologist who wishes to use chemical reagents should be extremely careful. In most cases the reagents are corrosive and often highly toxic. Avoid all contact with them and never eat the specimen on which the test has been carried out.

If these substances act in a visible and evident way which does not require the use of a microscope, they are called *macrochemical* reagents. We shall now deal briefly with the reagents most commonly used and most easily found; these will be some help in the identification of fungus species:
Strong bases Fairly concentrated solutions with 10% sodium

Right: The spores of puffballs are scattered in the atmosphere. This photo shows a handsome specimen of Geastrum fimbriatum (362), an earth star. Below: The spores of Laccaria laccata (76) are deposited on the grass beneath the cap, and turn it white.

carbonate (NaOH) and potassium hydroxide (KOH); particularly useful for the *Cortinarius* group, these are widely used in the practical systematics of fungi, as is ammonia (NH_4OH), both in solution and in vapor form.

Iron salts Ferrous sulphate ($FeSO_4$) in crystals or in water solution (10%) is commonly used for the *Russula* group, while ferrous chloride (Fe_2Cl_6) is useful for the *Cortinarius* group.

T1.4 (Henry's) reagent This reagent is used essentially for the *Cortinarius* group but also gives interesting reactions with other fungi. In preparing it great care must be taken because of the dangerous nature of the ingredients; if possible, have it prepared by specialized staff. The thallium oxide (1 g) is dissolved in nitric acid (4 cc) and concentrated hydrochloric acid (4 cc). Using great care, 1 g. of sodium bicarbonate is then added bit by bit to the solution.

Phenol or carbolic acid In 2% water solution it reacts in time with the flesh of almost all fungi, but a speedy reaction (about 1 minute) is a typical and specific characteristic of certain fungi.

Aniline Oil of aniline passed over the cuticle makes it possible to divide the field mushrooms (*Agaricus*) into two major groups, depending on whether they turn yellow or not. The trace of aniline crossed with concentrated nitric acid can produce a major reaction, colored orange-yellow, known as the

Method of determining and observing the spore color by putting the cap on a sheet of paper: light spores will be visible on a dark background, and dark spores on a light background.

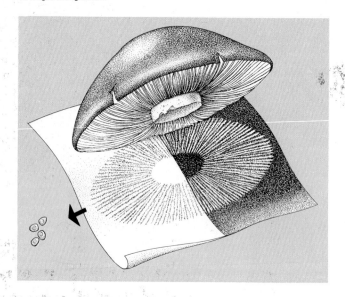

"Schaeffer's cross reaction." By stirring a few drops of oil of aniline in water one obtains water of aniline, an important reagent for the *Russula* group.

Phenolaniline Used for the *Cortinarius* group, it is prepared by putting three drops of oil of aniline into five drops of concentrated sulfuric acid and mixing these two with 10 cc of phenol at 2%.

Lygol An iodized water solution (1 g iodine, 2 g potassium iodide dissolved in 150–200 cc distilled water), this is sometimes used instead of the dangerous T1.4 reagent, but is used mainly for identifying fungi with amyloid flesh or hyphal elements, i.e. flesh which reacts blue-violet with amide.

Phenoloxidase reagents Not always specific, these give colored reactions with the flesh or fungi. The most widely used are: Guaiacol (guaiacum resin in water)—reacts red; Guaiacum tincture (guaiacum resin in alcohol, 60–70%)—blue; Pyramidon (solution in water) violet; Naphthol (solution in alcohol,) 30% gray-violet; Tyrosinem—first reddish, then black.

Microchemical reagents

Microchemical reagents often enable one to examine and identify mushrooms that otherwise are too difficult to name, mostly because of the large number of look-alikes in some groups. The most important reagents are:

Melzer's Reagent This is a yellowish compound that when ap-

Representations of some of the many different spore shapes.

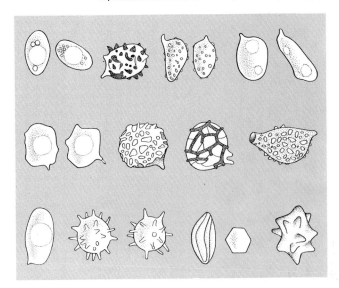

plied to mushroom tissue, spores in particular, may produce a blue-black (amyloid) or red-brown (dextrinoid) reaction, or none at all. This is useful in studying agarics, especially species of *Amanita* (some of which have amyloid spores) and *Lepiota* (most of which have dextrinoid spores). All species of *Lactarius* and *Russula* have amyloid ornamented spores, as do some other gill mushrooms. Some cup fungi have asci that are amyloid, especially around the tip, and this characteristic is used in differentiating some look-alike genera. Formula: Mix 20 g chloral hydrate, 0.5 g iodine, 1.5 g potassium iodide, and 20 cc water; warm this solution and stir.

Bases A weak solution (3–5%) of ammonium hydroxide (NH_4OH) or potassium hydroxide (KOH) is used to inflate mushroom tissues when preparing microscopic mounts of dried material. Either is also used to show staining of internal organs, such as cystidia (sterile cells) in *Naematoloma* and *Stropharia*, which become golden yellow in part.

Congo Red A 1% solution is used to stain the walls of hyphae.

Phloxine A 1% solution is used to stain the interior of hyphae.

Cotton Blue This stain is used to test the spore wall of agarics; those that become blue or darken appreciably are said to be cyanophilic. To prepare this stain, combine 50 cc of 1% solution of Cotton Blue in lactic acid (100 g), phenol (100 g), glycerine (150 cc), and 50 cc water.

THE KEYS

How to use them

Keys used in botany and zoology are groupings made on the basis of the most conspicuous characteristics of the organism in question, the aim being to make it as quick and easy as possible to identify the genus or species. The procedure in mycology is as follows:

Let us imagine that we have picked a mushroom which we do not know. Let us have a look at it: it is a fungus with a cap and a stipe; beneath the cap are pores. Now let us look at the corresponding key. The first question we must ask is: ''Can the tubes which run into the pores be detached from the cap or not?''

Let us try with a fingernail. Yes, they can. Across from ''detachable tubes'' is the number 2. Number 2 is therefore the next question we must answer. ''Is the stipe central and does the fungus grow on the ground?'' or ''Is the stipe lateral, is the fungus red, and does it grow on wood?'' In this case the stipe is central and it has been picked on the ground: a bolete.

This method of classifying fungi is not strictly ''scientific,'' but it enables us to find our way quite efficiently in the world of fungi. The characteristics of the carpophores examined progressively and alternately enables us to draw up the appropriate keys, which are a vital factor for anyone wishing to delve deeper into the study of fungi.

The keys which we include below have been considerably simplified and do not take microscopic characteristics into account. But despite the gaps inherent in them they will enable us to identify the most common genera.

Carpophore forms

1. with stipe and cap
with gills beneath
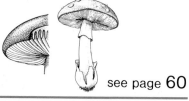
see page **60**

2. with stipe and cap
with pores beneath

see page **64**

3. with stipe and cap
with teeth beneath
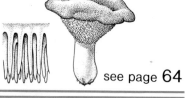
see page **64**

4. with stipe and hood-
shaped cap

see page **64**

5. bracket-shaped
with pores, gills,
or teeth beneath,
or smooth, on wood

see page **65**

6. cup-shaped, with or without stipe 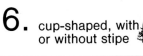  see page **66**

7. globose, pear-, or star-shaped   see page **66**

8. club-shaped or branched 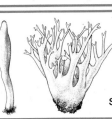 see page **68**

9. crustlike, smooth or with pores or teeth, on wood  see page **68**

10. gelatinous, mainly on wood see page **69**

1 FUNGI WITH STIPE AND CAP WITH GILLS BENEATH

a) spores white, cream or yellowish

1a	lignicolous (on wood)	2
b	terricolous (terrestrial)	12
2a	with eccentric to lateral stipe or none	3
b	with central stipe	8
3a	gill-edge entire	4
b	gill-edge serrated or split lengthwise	6
4a	cap fan-shaped, stipe short and thin, spores amyloid	**Panellus**
b	stipe large, often with several caps	5
5a	brown, leathery, dry	**Panus**
b	gray, bluish, brown, whitish, fleshy	**Pleurotus**
6a	edge of gill split, cap grayish, tomentose (downy)	**Schizophyllum**
b	gills serrate	7
7a	stipe eccentric, cap coarsely squamose, spores nonamyloid	**Lentinus**
b	stipe lateral to absent, cap smooth to hairy, spores amyloid	**Lentinellus**
8a	entirely white, slimy, with ring	**Oudemansiella mucida**
b	otherwise	9
9a	tufted, caps yellowish brown, with hairs, gills adnate to decurrent, with ring	**Armillariella**
b	tufted, caps orange, without hairs, gills decurrent, without ring	**Omphalotus**
c	otherwise	10
10a	gills notched	**Tricholomopsis**
b	otherwise	11
11a	fruit-body small, stipe often horsehairlike, long, slender, cap convex	**Marasmius** (in part)
b	stipe thin, fragile, cap conical to bell-shaped	**Mycena** (in part)
12a	gills adnate to decurrent	13
b	otherwise	17
13a	flesh brittle	14

b	flesh of stipe fibrous	15
14a	with latex (milk)	*Lactarius*
b	without latex	*Russula*
15a	gills crowded, thin-edged cap often funnel-shaped	*Clitocybe*
b	otherwise	16
16a	gills, distant, wedge-shaped, thin-edged, waxy	*Hygrophorus*
b	gills often distant, thick-edged, forked, or vein-like	*Cantharellus*
17a	with distant volva in young stage, often with ring, cap varied, gills typically free	*Amanita*
b	without volva, gills free or attached	18
18a	with ring on stipe or veil remains	19
b	without remains of partial veil	24
19a	cap viscous or glutinous in damp weather	*Limacella*
b	cap coarsely squamose, granular, dry	20
20a	cap finely granular	21
b	cap coarsely squamose, fibrillose or smooth	22
21a	fruit-bodies reddish or whitish, gills attached	*Cystoderma*
b	otherwise	*Lepiota* (in part)
22a	gills free	*Lepiota* (in part)
b	gills attached	23
23a	stipe with marginate bulb	*Cortinellus bulbiger*
b	stipe cylindrical, gills notched	*Tricholoma* (in part)
24a	spores smooth nonamyloid	*Tricholoma* (in part)
b	spores warty, amyloid	*Melanoleuca*
c	otherwise	25
25a	stipe thin and rubbery, often tufted, revives in water	*Marasmius* (in part)
b	stipe cartilaginous, nonreviviscent, cap convex	*Collybia*
c	stipe fragile, hollow, non-reviviscent, cap bell-shaped	*Mycena* (in part)

b) **spores pink to salmon-colored**

1a	stipe lateral or absent	2
b	stipe central to eccentric	3
2a	fruitbodies large, spores pinkish	*Phyllotopsis*
b	frúitbodies small, spores brownish	*Crepidotus*
3a	with volva	*Volvariella*
b	without volva	4
4a	gills free, typically on wood	*Pluteus*
b	gills attached, typically on ground	5
5a	gills adnexed to notched	6
b	gills long decurrent, spores with longitudinal ridges	*Clitopilus*
6a	spores angular, pinkish salmon	*Entoloma*
b	spores smooth to minutely roughened, pinkish buff	*Clitocybe (Lepista)*

꞊) **spores ochreous or brown**

1a	stipe lateral, fruit-body fan-shaped, on wood	*Crepidotus*
b	otherwise	2
2a	with veil or membranous ring	3
b	without veil or ring	7
3a	large terrestrial fungus with granular cuticle, tawny-yellow	*Phaeolepiota aurea*
b	cap smooth, white or brown, stipe smooth, not scaly	*Agrocybe*
c	cap typically scaly	*Pholiota*
d	fruit-bodies thin, on moss	*Galerina*
e	otherwise	4
4a	on wood, large, orange-yellow with ring	*Gymnopilus spectabilis*
b	otherwise	5
5a	quite variable in color and shape, rust-brown spores, veil cobwebby	*Cortinarius*
b	cap dry, fibrillose, coarsely squamose, spermatic or other odor	*Inocybe*
꞊	cap viscid, stipe rooting, odor of bitter almonds	*Myxocybe radicata*

	d	otherwise	6
	6a	cap fleshy, yellow or red-dish	**Gymnopilus** (in part)
	b	small species, brownish, reviviscent	**Phaeomarasmius**
	7a	gills decurrent, margin involute	**Paxillus**
	b	otherwise	8
	8a	radishy odor, gills adnate, cap slimy, sometimes with veil when young	**Hebeloma**
	b	otherwise	9
	9a	cap shiny yellow, fragile	**Bolbitius**
	b	otherwise	10
	10a	cap pointed and viscid, with stipe rooting	**Phaeocollybia**
	b	cap acutely conical, moist to dry, stipe not rooting	**Conocybe**
d)	**spores purplish brown or blackish**		
	1a	gills self-digesting, i.e. dissolving into a sort of ink	**Coprinus**
	b	gills not self-digesting	2
	2a	gills decurrent, with glutinous veil	**Gomphidius**
	b	otherwise	3
	3a	cap dry, gills free or attached	4
	b	cap glutinous-viscid or tacky, gills attached	5
	4a	cap white or brown, sometimes coarsely squamose, with ring, gills free	**Agaricus**
	b	with or without ring, tufted, thinner, gills attached	**Psathyrella**
	5a	often yellowish with cobweblike veil, typically tufted	**Naematoloma**
	b	otherwise	7
	6a	ring or remains visible	**Stropharia**
	b	without ring and/or bruising blue	**Psilocybe**

2 FUNGI WITH STIPE AND CAP WITH PORES BENEATH

1a	tubes detachable	2
b	tubes not detachable	3
2a	stipe central, typically terrestrial	*Boletus*
b	stipe lateral, liver-red, on wood	*Fistulina*
3a	on wood	4
b	terrestrial	*Polyporus* (in part)
4a	stipe central	*Polyporus* (in part)
b	stipe lateral	5
5a	surface crusty, shiny	*Ganoderma*
b	without crust, flesh brown	6
6a	spores white	*Phaeolus*
b	spores yellowish	*Coltricia*

3 FUNGI WITH STIPE AND CAP WITH TEETH BENEATH

1a	stipe central	2
b	stipe lateral	4
2a	flesh pale, spores cream-colored	*Dentinum*
b	flesh brightly colored, often zonate	3
3a	consistency fleshy, tough, often bitter taste	*Hydnum, Hydnellum*
b	consistency tough or leathery	*Phellodon*
4a	gelatinous, gray	*Pseudohydnum*
b	otherwise	5
5a	on pinecones, surface velvety, stipe present	*Auriscalpium*
b	otherwise	6
6a	base bulbous, club-shaped, with long teeth, on wood	*Hericium* (in part)
b	teeth erect and divergent, all over surface, on wood	*Hericium* (in part)

4 FUNGI WITH STIPE AND HOOD-SHAPED CAP

1a	honeycomblike cap	*Morchella*
b	smooth or folded	2
2a	cap smooth or folded, bell-shaped	*Verpa*
b	cap with contorted folds, brain-shaped	*Gyromitra*
c	otherwise	3
3a	cap saddle- or cup-	

	shaped	*Helvella* (in part)
b	cap knoblike, irregular	4
4a	cap gelantinous with smooth base	*Leotia*
b	cap dry, with tomentose base	*Cudonia*

5 BRACKET-SHAPED FUNGI

1a	with gills	*Lenzites*
b	with teeth	2
c	with pores	3
2a	with completely free teeth	*Irpex*
b	with teeth joined at the base	*Sistotrema*
3a	tubes not forming distinct layer	4
b	hymenial layer distinct from flesh	6
4a	with mazelike hymenium	*Daedalea*
b	otherwise	5
5a	pores very large, hexagonal, surface crustlike	*Hexagona*
b	*pores average to small, roundish, surface hairy*	*Trametes*
6a	flesh basically tender, also fragile when hardened	7
b	flesh hard to ligneous	8
7a	cap spongy, bristly	*Spongipellis*
b	surface smooth	*Tyromyces*
8a	surface crustlike	9
b	surface velvety, not crustlike when young, flesh yellow-brown	11
9a	typically applanate, surface lacquered	*Ganoderma*
b	otherwise	10
10a	applanate to hoof-shaped, flesh and pores whitish	*Fomitopsis*
b	hoof-shaped, flesh and pores brownish	*Fomes*
c	hoof-shaped, flesh and pores yellow-brown to orange-brown	*Phellinus*
11a	large with short thick stalk, whitish spores	*Phaeolus*
b	small with slender stalk, yellowish spores	*Coltricia*

6 CUP- OR SAUCER-SHAPED FUNGI

1a	bright red, yellow-green, or violet	2
b	brown or blackish	8
2a	terrestrial	3
b	on wood	4
3a	red with smooth edge, or ochre-yellow	***Peziza*** (in part)
b	red with hairy edge	***Scutellinia***
4a	red with pubescent outer surface	***Sarcoscypha***
b	otherwise	5
5a	green, on green wood	***Chlorosplenium***
b	otherwise	6
6a	violet	***Ascocoryne***
b	yellow	7
7a	yellow, smooth	***Bisporella***
b	yellow or whitish, hairy	***Dasyscyphus***
8a	brown, on wood, fir cones, chestnut husks, etc.	***Rutstroemia***
b	otherwise	9
9a	brown with hypogeal sclerotia	***Sclerotinia***
b	otherwise	10
10a	cup with grooved stipe	11
b	otherwise	12
11a	stipe slender	***Helvella*** (in part)
b	stipe thick	***Paxina***
12a	reddish brown, flattened, with rhizoids from lower surface	***Rhizina***
b	without rhizoids	***Peziza*** (in part)

7 GLOBOSE, PEAR- OR STAR-SHAPED FUNGI

1a	on earth or wood (epigeal)	2
b	subterranean (hypogeal)	11
2a	gleba (spore mass) powdery at maturity	3
b	gleba not powdery	8
3a	fruit-body opens in form of a star, with globose center, with or without ostiole	***Geastrum***
b	central part pierced irregularly, with fruit-body opening and closing depending on moisture	***Astreus***
c	central part with many ostioles	***Myriostoma***

d	fruit-body not opening star-like	4
4a	globose part with peduncle normally sunk in the ground	*Tulostoma*
b	stipe short or absent	5
5a	outer layer yellowish, thick, mature gleba blackish	*Scleroderma*
b	outer layer smooth, with scales or warts, slender, white, gray, or brown	6
6a	with stipe, gleba with blackish blue spherules, in sandy ground	*Pisolithus*
b	otherwise	7
7a	peridium smooth, consisting of two layers, the outer one dropping off at maturity, the inner turning gray or blackish, with no sterile part	*Bovista*
b	peridium smooth, just one layer, no sterile part	*Langermannia (Calvatia gigantea)*
c	external peridium with spines, sterile tissue at base	*Lycoperdon*
8a	on wood	9
b	terrestrial	10
9a	blackish, concentrically zoned internally	*Daldinia*
b	reddish brown, blackish, small	*Hypoxylon*
c	vermilion-red	*Nectria*
10a	white, with a gelatinous layer inside and red or green gleba	*Phallus, Clathrus, Anthurus,* etc.
b	with no inner gelatinous layer	11
11a	yellowish or reddening peridium, gleba with distinct cells	*Rhizopogon*
b	otherwise	12
12a	finely warty peridium, yellowish brown, gleba first slightly venose, then powdery	*Elaphomyces*
b	peridium finely warty or smooth, gleba venose	*Tuber* (in part)
c	peridium black, markedly warty	*Tuber* (in part)

8 CLUB-SHAPED OR BRANCHED (BUSHLIKE) FUNGI

1a	club-shaped or slightly branched	2	
b	plentifully branched	6	
2a	on wood	3	
b	on other matter	4	
3a	black, leathery	*Xylaria*	
b	yellow, elastic, slightly viscous	*Calocera*	
4a	finely speckled or with warts, on insects or hypogeal fungi	*Cordyceps*	
b	terrestrial	5	
5a	yellow, smooth	*Spathularia*	
b	black or greenish	*Geoglossum*	
c	yellow, yellowish, white	*Clavaria*	
6a	dark brown, leathery with slightly flattened ramifications, terrestrial	*Thelephora*	
b	otherwise	7	
7a	on wood, yellow, tough to horny	*Calocera*	
b	on wood, black, leathery	*Xylaria*	
c	otherwise	8	
8a	markedly ramified, with flattened branches	9	
b	markedly ramified with cylindrical branches	10	
9a	with smooth hymenium	*Sparassis*	
b	with porous hymenium	*Grifola*	
10a	thick branches	*Ramaria*	
b	slender, threadlike branches	*Pterula*	

9 CRUSTLIKE FUNGI ON WOOD

1a	crustlike with raised margins	2	
b	completely crustlike	8	
2a	with pores or gills	3	
b	reticulate or with teeth	5	
3a	with gills	4	
b	round or rhomboid pores	various polypores	
4a	flesh brown	*Lenzites*	
b	flesh white, often on oak	*Daedalea*	
5a	with teeth	6	
b	reticulate	7	
6a	gray, gelatinous	*Pseudohydnum*	
b	yellowish, fragile	*Steccherinum*	
7a	gelatinous, orange, radiating folds, white spores	*Merulius*	

b	spongy, yellow-brown radiating folds, brown spores	*Serpula*
8a	blackish	9
b	otherwise	10
9a	unevenly crustlike	*Ustulina*
b	disclike	*Diatrype*
10a	orange	*Phlebia*
b	white, grayish or brown	11
11a	with small warts, olive-brown	*Coniophora*
b	smooth, brownish, sometimes purple	*Stereum*

10 GELATINOUS FUNGI

1a	with toothed hymenium and stipe	*Pseudohydnum*
b	otherwise	2
2a	brown, yellow, circumvolute	3
b	pendant- or cup-shaped	4
3a	yellow, on living juniper	*Gymnosporangium*
b	yellow or brown, on dead wood	*Tremella*
c	blackish	*Exidia*
4a	pendant-shaped, yellowish, on conifers	*Dacrymyces*
b	other colors	5
5a	ear-shaped, with violet-brown hymenium	*Auricularia*
b	otherwise	see Cup- or Saucer-shaped Fungi

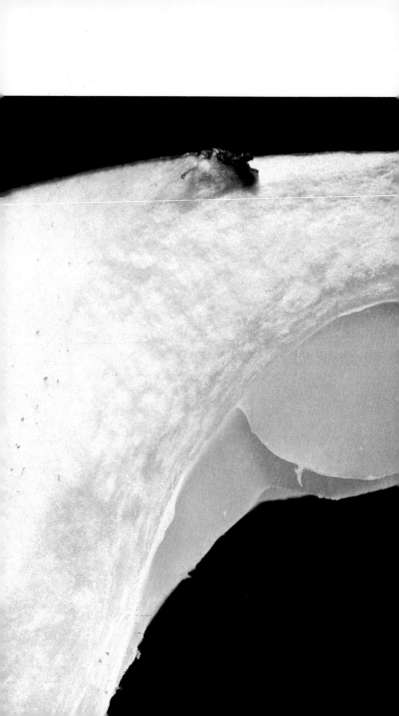

1 AMANITA CAESAREA

Etymology From Latin, "of Caesar" or "regal," because it was a favorite of the early Roman Caesars.

Description Cap 8–20 cm, hemispherical to flat, orange-red washing out to yellow, cuticle separable, sometimes with evident membranous remains of white veil, margin striate. Gills free, crowded, yellow. Stipe 8–15 × 2–3 cm, narrowing at top, hollow when mature, with yellow falling ring, slightly swollen at base, with large white membranous volva. Flesh whitish, yellowish beneath cuticle. Without evident odor. Spores white, elliptical, smooth, 8–14 × 5–8.5 microns.

Edibility In Europe, excellent cooked.

Habitat In airy parts of dry oak woods, in slightly acid ground, and with pines.

Season Spring to autumn.

Note The common North American form, which may be a distinct species, has a somewhat umbonate cap and a thinner (1–2 cm) yellow stripe, and occurs from eastern Canada to Florida and west to the central states; the same or a similar form occurs in the Southwest and Mexico.

Caution A number of look-alikes of unknown edibility exist in North America. And what is called *A. caesarea* in North America is not generally regarded as choice.

2 AMANITA MUSCARIA
Muscaria

Common name Fly agaric.

Etymology From Latin "of flies," because of the northern European custom of using the cap, soaked in milk, to kill or stupefy flies.

Description Cap 8–25 cm, hemispherical to slightly concave, cuticle detachable, red, covered with white pyramidal warts which may be removed by rain, margin striate. Gills white, crowded, free. Stipe 12–25 × 1.5–2.5 cm, basal bulb with volva of several concentric warty rings, ring white, membranous. Flesh soft and white, orange-red beneath cuticle. Not very conspicuous odor. Spores white, ovoid, smooth, 9–11 × 6–8 microns.

Edibility Fairly poisonous, depending on the season.

Habitat In mountains under conifers and birch.

Season Summer and fall.

Note Although reported from Siberia as producing hallucinations, the red-capped North American variety which is found in northern forests and higher altitudes in the south, causes delirium, manic behavior and deep sleep, sometimes accompanied by profuse sweating.

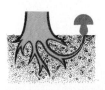

3 AMANITA MUSCARIA
Formosa

Common name   American fly agaric.
Etymology   From Latin, "handsome-looking."
Description   This variety has a yellowish to orange coloration, or has a tinge of red at the center. All the velar remains (ring, volva and warts) are white, and the description of *A. muscaria* (**2**) applies to this variety in all other respects.
Edibility   Toxicity appears to vary widely from place to place, and seems to lie for the most part in the cap cuticle.
Habitat   This variety is very common in North America, but becomes rarer, more slender, and tinged with a salmonlike coloration in the southern states. Moving west we also find the typical red-capped *A. muscaria*.
Season   Summer and autumn.
Note   Although the toxins in both varieties of this species are reportedly concentrated in the colored skin of the cap, peeling the mushroom does not render it harmless, and poisonings do occur. The hallucinations for which the red-capped Siberian variety is notorious do not seem to occur with either American variety; rather, the experience is often one of delirium and deep sleep, sometimes accompanied by profuse sweating.

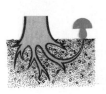

4 AMANITA PHALLOIDES

Common name   Death cap.
Etymology   From Latin, "phalluslike," because of the shape of the carpophore in the early stages of growth.
Description   Cap 5–20 cm, subspherical to flat, rarely with membranous velar remains; fairly deep olive-green to olive-brown but paler toward margin, usually with dark innate radial fibrils. Gills white or slightly yellowish, quite crowded, free. Stipe 8–20 × 1–2 cm, tapering toward top, hollow when mature, white speckled with greenish gray stripes, white membranous ring, striate at top, base bulbous with large, white, membranous volva. Flesh white, but greenish yellow just beneath cuticle. Odor first neutral, then nauseous. Spores white, ovoid to nearly round, smooth, 8–11 × 7–9 microns, amyloid.
Edibility   Deadly poisonous.
Habitat   With a preference for broadleaf trees, particularly oak, but also under pine and spruce.
Season   Late spring to late autumn.
Note   When cut, the outline of the carpophore is white, faintly green in the cuticle area. As little as one cap can prove fatal to an adult. Symptoms don't occur for about 10–12 hours (or longer). Although only recently confirmed in North America, it is now known from several eastern seaboard states and the Pacific coast.

5 AMANITA CITRINA

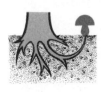

Etymology From Latin, "lemonlike," because of the color of the cap.

Description Cap 6–12 cm, hemispherical to flat, cuticle various shades of yellow, covered with flat, irregular warts, whitish turning to brown, margin smooth. Gills crowded, whitish, free, sometimes semifree. Stipe 6–12 × 0.5–1.5 cm, whitish cap-colored with membranous rings, bulbous base, whitish turning to brown, with a large, punky, marginate volva. Flesh white, faintly yellow beneath cuticle. Odor of radish or potato. Spores white, subglobose, smooth, 7–10 × 6–9 microns, amyloid.

Edibility Slightly poisonous when raw, rather mediocre cooked.

Habitat In fine, sandy soil, preferably beneath broadleaf species.

Season Summer and autumn.

Note The presence of an irritant substance like that secreted by the skin glands of toads makes this mushroom slightly poisonous when eaten raw.

Caution *A. citrina* should not be eaten either raw or cooked because its cap color is often difficult to distinguish from that of the deadly *A. phalloides* (**4**).

6 AMANITA GEMMATA

Etymology From Latin, "bejeweled," because of the warts on the cap.

Description Cap 4–8 cm, hemispherical to flat, cuticle slightly sticky, straw- or orange-yellow, sometimes with floccose volva fragments, easily detachable, margin finely striate. Gills free, semifree, or slightly adnate, white. Stipe 6–10 × 1–2 cm, narrowing toward top, full when mature, whitish, with fragile short-lived ring, volva close to basal bulb which may split into one or two rings or collars. Flesh tender, white, slightly lemon-yellow beneath cuticle. No odor. Spores white, elliptical, smooth, 10–12 × 7–8 microns.

Edibility Edible, although it has apparently caused cases of poisoning.

Habitat Beneath broadleaf and coniferous trees, with a preference for acid soil.

Season Late autumn to spring; year-round where suitable climatic conditions exist.

Note This species is fairly variable. In the Pacific Northwest this species seems to intergrade (hybridize) with the poisonous *A. pantherina* (**10**), and is difficult to identify and dangerous to eat.

Caution In the Northeast, where it is eaten by some, there have been reported poisonings.

7 AMANITA VERNA

Common names Fool's mushroom; spring destroying angel.
Etymology From Latin, "spring," which is when it appears.
Description Cap 3–10 cm, hemispherical to flat, cuticle white, silken, sometimes with shades of ochre at center, normally without universal velar remains, margin thin and smooth. Gills white, crowded, and free. Stipe 7–13 × 1–1.5 cm, narrowing at top, hollow when mature, white with ring and membranous saclike volva, both fairly close, base sometimes slightly enlarged. Flesh white. No odor in young specimens, pungent in mature ones. Spores white, subspherical, smooth, 8–10 × 7–9 microns, amyloid.
Edibility Deadly poisonous.
Habitat Preference for broadleaf trees in fine, sandy, acidic soil, in warm temperate regions.
Season Spring and summer, also found in autumn.
Note Unlike many poisionous plants, which are bitter or otherwise unpalatable, some poisonous mushrooms, and especially the deadly *Amanitas*, taste good. It takes 8–12 hours or more before the appearance of any symptoms.

8 AMANITA VIROSA

Common name Destroying angel.
Etymology From Latin, "fetid," because of its smell.
Description Cap 4–10 cm, conical to convex, never completely opened, white. Cuticle detachable, margin smooth. Gills white, crowded, free with pruinose aspect. Stipe 8–15 × 1–1.5 cm, narrowing at top, hollow when mature, with wooly-fibrillose surface, ring incomplete and fragile, volva membranous, saclike and close to stipe. Flesh white, soft. Usually an unpleasant, yeastlike odor. Spores white, round, smooth, 9–12 microns, amyloid.
Edibility Deadly poisonous.
Habitat In sandy, acid soils, with a preference for broadleaf trees in mountainous regions or in cold temperate climates.
Season Spring through autumn.
Note The broadleaf woods of North America play host to *A. bisporigera*, also deadly poisonous, and similar in appearance and color to *A. virosa*, but with a more slender and smooth stipe. The cuticle of both species gives a golden yellow reaction with KOH, a reaction lacking in *A. verna*. *A. ocreata*, a late winter mushroom in central and southern California, is equally deadly, and differs in that its cap discolors pinkish ochre about the center as it ages.

9 AMANITA SPISSA

Etymology From Latin, "massive" or "huge," because of its appearance.

Description Cap 7–15 cm, hemispherical to flat, gray-brown with grayish white warts, margin smooth. Gills white, crowded, free. Stipe 8–12 × 1–2.5 cm, narrowing toward the top, full, with a white, membranous ring, striate like the upper part of the stipe, small, close, brown scales beneath, basal bulb (or napiform base), volva small. Flesh white, constant, firm. Slightly radishlike odor. Spores white, ovoid, smooth, 8–10 × 6–8 microns, amyloid.

Edibility Good in Europe. **Avoid in North America.**

Habitat In broadleaf and coniferous woods.

Season Summer and autumn.

Note This species, which is eaten in Europe, should not be eaten in North America because it resembles the poisonous *A. pantherina* and its form and distribution are not known well enough for reliable identification.

Caution See note.

---

10 AMANITA PANTHERINA

Common name Panther.

Etymology From Latin, "panther," because of the spotted appearance of the cap.

Description Cap 6–12 cm, hemispherical then flat, cuticle brown covered with small white pyramidal warts, short-lived, margin clearly striate. Gills white, crowded, free. Stipe 6–12 × 0.5–2 cm, tapering toward the top, with membranous white ring with striate surface toward the gills and basal bulb with a marginate volva often split into two or three rings. Flesh white, not very firm. Odor at first neutral, then acrid and nauseous, flavor quite sweet. Spores white, ovoid, smooth, 10–12 × 7–8 microns.

Edibility Poisonous, symptoms similar to those produced by *A. muscaria* (2), but more serious.

Habitat Beneath broadleaf and coniferous species.

Season Summer and autumn.

Note The cap color varies from dark brown to yellowish or cream to nearly white, and its best identifying characteristics are its volva, striate cap margin, and typically unchanging white flesh. It is common in the Rocky Mountains and Pacific Northwest.

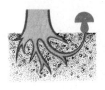

11 AMANITA RUBESCENS

Common name Blusher.

Etymology From Latin, "reddening," because of the discoloration of the flesh on bruising or aging.

Description Cap 5–20 cm, semiglobose to flat, cuticle reddish brown to whitish, with wine-red markings, covered with grayish pink warts, easily washed out by rain, margin smooth. Gills free, crowded, white, with wine-red markings particularly after being picked. Stipe 7–20 × 1–3 cm, fairly sturdy at first, tapering toward top, hollow when mature, tawny to brown, with coppery shades, large membranous ring, white often speckled with reddish brown markings, especially at edge, basal bulb with scattered scale-shaped universal velar remains. Flesh white, slowly turning to red when exposed to air, especially stipe. Odorless. Spores white, ovid, smooth, 8–10 × 6–7 microns, amyloid.

Edibility Excellent cooked, fried, **toxic raw.**

Habitat In broadleaf and coniferous environments, on any dry ground.

Season Spring to late autumn.

Note If for eating take care not to collect *A. pantherina* (10) or, in the Northeast, *A. brumescens* (which has a conspicuously split, large basal bulb).

Caution Toxic when raw. See note.

---

12 AMANITA SOLITARIA

Etymology From Latin, "solitary," because generally isolated specimens are found.

Description Cap 5–20 cm, hemispherical to flat, cuticle moist and whitish, removable, with large grayish warts, margin smooth, large velar remains. Gills white, crowded, free or slightly adnate. Stipe 6–25 × 1.5–3.5 cm, white, sturdy, always solid, cottony, with short-lived ring, mealy velar remains. Flesh white, soft. Odor insignificant. Spores white, elliptical, smooth, 10–13 × 7–9 microns, amyloid.

Edibility Good in Europe, better when earthy-tasting cuticle is removed. **Avoid in North America.**

Habitat In sunny areas in broadleaf woods, on calcareous ground.

Season Summer through early autumn.

Note Similar to *A. solitaria* (which has not been verified in North America) is *A. smithiana* of the Pacific Northwest, which is toxic and to be avoided. Various eastern varieties (sometimes smelling of chlorine) cannot be identified without a microscope, so avoid all American mushrooms that look like *A. solitaria*.

Caution Avoid in North America. See note.

13   AMANITA ECHINOCEPHALA

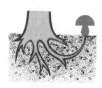

Etymology From Greek, "spiny-headed."

Description Cap 7–20 cm, subglobose, hemispherical, convex, flat or slightly depressed, whitish tending to become ochre-colored, covered with large pyramidal warts (usually pointed), on the whole adherent then falling off from the edge inward, margin with cottony remains of partial veil, cuticle completely removable. Gills whitish, free or slightly adnate, with creamy cottony quality, crowded, with cream-yellow or pale green shading when mature. Stipe sturdy, 8–20 × 1.5–2.5 cm, often with bulb, tapering toward top and pointed at base, whitish, covered at base with scales of the same color produced by the ring. Ring large, membranous, striate lengthwise, floccose at the margin. Flesh off-white, soft, especially in the stipe, tending to pale yellow when drying. Slightly earthy odor, no significant flavor. Spores white, with green or yellow shades of color, elliptical, smooth, 9–12 × 8–11 microns, amyloid.

Edibility Nontoxic, but poor in flavor.

Habitat Beneath conifers or broadleaf species, in calcareous soil, in dry sunny places.

Season Late spring and summer, less common in autumn.

Note Not known to occur in North America, but all look-alikes should be avoided—some are poisonous.

14   AMANITA INAURATA

Etymology From Latin, "gilded," from the cap color.

Description Cap 8–12 cm, companulate to convex, with slight umbo if flat or depressed, varying from grayish brown to olive-ochre-yellow to yellowish brown and other shades. Darker at the disc, with paler marginal area, markedly striate, with charcoal-gray velar patches. Gills free, whitish, turning gray or cream-colored when mature. Stipe 12–20 × 1.5–3 cm, slightly narrower at top, no ring, with gray or brownish stripes, hollow, with white volva turning grayish, reduced to ring fragments at base which is only slightly enlarged. Flesh white. No odor. Spores white, spheroid, smooth, 12 × 14 microns.

Edibility Adequate. **Avoid in North America.**

Habitat Woodland; no special environmental requirements.

Season Summer and autumn.

Note The group of ringless *Amanita* species has many members with distinctive features which overlap in the intermediate forms, and are thus not easily identifiable. Because of the difficulty encountered in identification, mushrooms so identified should not be eaten: not enough is known about North American varieties and look-alikes.

Caution Avoid in North America. See note.

15 AMANITA FULVA

Synonym *Amanita vaginata* v. *fulva.*
Common name Tawny grisette.
Etymology From Latin, "reddish" or "tawny," after the color.

Description Cap 4–10 cm, conical to flat, reddish yellow, glabrous, markedly striate margin. Gills white, free, long, crowded. Stipe 8–12 × 0.5–1.5 cm, narrowing at top, whitish or uniformly pale orange, ringless, volva whitish, membranous, sheathlike. Flesh fragile and white. Odor neutral. Spores white, spherical, smooth, 8–12 microns.
Edibility Mediocre.
Habitat In acid ground, under broadleaf species, often among rotting stumps.
Season Summer and early autumn.

Note A stipe with orange scales and the orange interior of the volva are the distinguishing features of *A. crocea,* a similar though less common species mainly found in broadleaf woods in warm temperate regions.

16 AMANITA UMBRINOLUTEA

Etymology From Latin, "yellowish brown," after the color of the cap.
Description Cap 4–10 cm, conical then flattened with a central umbo, color varying from brown to ochre-gray, margin markedly striate, with darker zonation at start of striations. Gills crowded, long, free, and white. Stipe 8–17 × 0.5–2 cm, narrowing at top, tapered, hollow when mature, white with orange-brown stripes, ringless, volva membranous, tall, sheathlike, whitish. Flesh fragile, subtle, white. Odorless and without specific flavor. Spores white, elliptical, ovoid, smooth, 11–16 × 5–13 microns.

Edibility Good.
Habitat Preference for beech and fir woods, often in open places.
Season Summer through autumn.

Note In the very large group of species which resemble *A. vaginata,* a ringless species with striate cap margin, this one is easily recognizable because of the darker circular area before the cap striations, and because of the spores, which are not round.

17 LIMACELLA GUTTATA

Synonym *Limacella lenticularis; Lepiota lenticularis.*
Etymology From Latin, "with droplets."
Description Cap 4–12 cm, subglobose to convex-campanulate, pinkish cream-colored, deeper at disc, paler at margin, which is slightly glutinous in damp weather, otherwise smooth and glabrous. Gills whitish, sometimes with olive-green shading, free, crowded, ventricose, often forked. Stipe 8–10 × 1–2 cm, white or cream-colored, cylindrical or slightly enlarged at the base, solid, the upper part covered with droplets of moisture in very damp or wet weather. Ring membranous, pendant, whitish, marked like the stipe. Flesh white, faintly reddish near base, yellowish at base. Strong odor of flour, flavor sweetish. Spores white, elliptical, smooth, 7–8 × 4–5 microns.
Edibility Good.

Habitat In damp woodland, mainly broadleaf, particularly beech, ash, and elm, but also beneath conifers.
Season Summer and autumn.
Note A relatively rare genus, *Limacella*, like *Amanita*, contains white-spored mushrooms (with free or nearly free gills) that develop from an egglike universal veil, but in *Limacella* this veil is slimy.

18 CYSTODERMA AMIANTHINUM

Etymology From Greek, "uncontaminated" or "pure."
Description Cap 3–5 cm, ochre-yellow tending to orange, quite fleshy, convex then flat with slight umbo at center, furfuraceous and granulose, often rugose. Gills cream-colored, adnate, crowded. Stipe 3–6 × 0.4–0.6 cm, whitish above ring, covered, below it, with ochre-colored granules, cylindrical. Ring ochreous, inferior, granulose like stipe. Flesh yellow, darker toward base. Loamy odor and sweetish flavor. Spores white, elliptical, smooth, typically 4.5–7.5 × 3–3.5 microns, amyloid.
Edibility Mediocre.

Habitat In coniferous woods, and on heathland and moors.
Season Summer and autumn.
Note *Cystoderma* is a genus of white-spored mushrooms that are generally orange in color and have a granular cap and stipe, with attached gills, and usually a ring. Most are difficult to identify without a microscope and chemicals. *C. amianthinum rugosoreticulatum* has a distinctly wrinkled cap, a strong green-corn odor, and is usually found in moss. *C. fallax* has a large, flaring, persistent ring on the stipe.

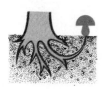

19 CYSTODERMA CARCHARIAS

Etymology From Greek, "with a rough surface."

Description Cap 2–5 cm, sometimes white but usually shaded with pinkish or, more rarely, pale lilac, convex, flat, often umbonate, covered with minute granules, with appendiculate margin or cap edge. Gills white, crowded, adnate. Stipe 3–6 × 0.4–0.8 cm, cap-colored below ring and covered with small, pointed warts, white higher up, slightly enlarged at base and slightly narrower at top. Ring of the same color, smooth on interior, like the lower part of the stipe externally. Flesh whitish or ochreous. Strong fetid smell and unpleasant flavor. Spores white, elliptical, smooth, 4–5 × 3–4 microns, amyloid.

Edibility Can be eaten though quality poor.

Habitat Coniferous woods, mossy meadows, and fields.

Season Summer and autumn.

Note It is not very clear whether *Cystoderma* is symbiotic or saprophytic. *C. ambrosii* is an entirely white species with the cap initially covered with small conical warts, becoming fairly smooth, as does the stipe. *C. ponderosum*, with the cap up to 11 cm, is ochreous or yellowish brown, scaly, with a whitish stipe and inconspicuous ring.

---

20 MACROLEPIOTA PROCERA

Synonym *Lepiota procera*.

Common name Parasol mushroom.

Etymology From the Latin, "tall," after its height.

Description Cap 10–25 cm or more, whitish with a thick brown cuticle which breaks up into large scales, short-lived and detachable from the edge inwards, remaining whole on the wide raised umbo. Mainly oval, then broadly campanulate turning to flat, with fibrillose, frayed edge. Gills whitish, turning darker, free, crowded, ventricose, wide, soft. Stipe 15–30 × 1.5–2 cm or larger, with markedly brown-striped surface, base with large flat bulb. Ring bandlike, movable. Flesh soft in cap, fibrous in stipe, whitish. Pleasant odor and hazelnutlike flavor. Spores white, elliptical, smooth, 15–20 × 10–13 microns, dextrinoid.

Edibility Cap excellent.

Habitat A common species in meadows and open woods.

Season Summer and autumn.

Note This mushroom is delicious, so keep in mind its salient features: long, slender, scaly stipe, scaly cap with an often prominent umbo, free gills, and ring which can be loosened and freely moved up and down the stipe.

21 MACROLEPIOTA RHACODES

Synonym *Lepiota rhacodes*
Common name Shaggy parasol.
Etymology From Greek, "frayed" or "ragged."
Description Cap 7.5–20 cm, globose then convex to flat, cuticle brownish, smooth, soon breaking into concentrically arranged polygonal scales, cap margin usually frayed. Gills whitish, bruising yellowish to brown, free, crowded. Stipe 10–20 × 1.5–2.5 cm, club-shaped, often with large basal bulb; double ring, frayed, movable, white, aging brownish. Inconspicuous odor and flavor. Spores 8–15 × 6–8 microns, white, elliptical, smooth, dextrinoid.
Edibility The caps are very good but see **note** below.
Habitat In cultivated fields, beneath trees in parks and meadows, and in open woods.
Season Summer and autumn.
Note Do not confuse this mushroom with the poisonous *Chlorophyllum molybdites,* which is common in grassy areas in southern states. It is very similar when young but is distinguishable as it matures: its gills turn a dirty grayish green and produce a greenish spore print.

22 MACROLEPIOTA PUELLARIS

Synonym *Lepiota puellaris.*
Etymology From Latin, "childlike," because of its pure white color.
Description Cap 4–8 cm, campanulate then convex, white, slightly raised at the ochre-colored disc, turning brown, with the surface breaking up into small floccose, raised scales, tending to be darker at the tip. Gills white, free, fairly crowded. Stipe 9–13 × 1–2 cm, enlarged at the base, almost bulbous, narrowing toward the top, hollow, slightly farinose above the ring, white. Ring white, movable. Flesh white, pinkish at edges of stipe, where it is also fibrous. Odorless and with no specific flavor. Spores white, elliptical, smooth, 12–18 × 7–8 microns, dextrinoid.
Edibility Good.
Habitat In meadows, among ferns, or near trees, or in sunny woods, prefers conifers, often in groups.
Season Spring to autumn.
Note The slight color change of the flesh may cause confusion with *M. rhacodes,* of which it has long been considered a variety. In southern European coastal pinewoods, where it also occurs in the winter, the mycelium embraces both sand and humus and resembles a foot with roots.

23 LEUCOAGARICUS PUDICUS

Etymology From Latin "chaste," from its pure color.

Description Cap 5–10 cm, white or whitish, with slight shades of pink or yellow, globose then spreading, with a hint of an umbo, cuticle thin giving the cap, when it breaks up, a granular or pubescent appearance, but never squamose. Gills white, free, ventricose, crowded, soft. Stipe 4–8 × 1–1.5 cm, white, narrowing toward the top, hollow when mature, fibrous, slightly enlarged at the base, with a white ring, ascending, with frayed margin, short-lived (dropping off) in old specimens. Flesh white, thick in the cap, soft. Odorless and without specific flavor. Spores white, slightly pinkish, ovoid, smooth, 7–9 × 4.5–6 microns, dextrinoid.

Edibility Young caps very good, but see **caution.**

Habitat In meadows, gardens, and grassy areas in woods.

Season Spring to autumn.

Note It resembles a field mushroom, *Agaricus,* but has white spores. *L. naucinus* is the common North American species. It has a smooth to minutely scaly white to sometimes grayish cap, and white gills aging pinkish.

Caution This ubiquitous, late summer and early fall lawn and park mushroom may cause digestive upsets; gray-capped or iodine-smelling forms should not be eaten.

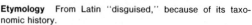

24 CYSTOLEPIOTA ADULTERINA

Etymology From Latin "disguised," because of its taxonomic history.

Description Cap 2–5 cm, semiglobose-campanulate, then spreading, mainly whitish, then grayish brown, rather rust-colored in old damp specimens, covered by thick farinose granulation. Gills whitish, tending to turn grayish. Stipe 4–6 × 0.3–0.6 cm, tapered, mealy like the cap, with a farinose mycelial envelope at base resembling a swelling, whitish then with rust-colored marking, veil with mealy consistency only visible in young specimens. Flesh whitish, tending to turn red in stipe toward bottom. No particular odor or flavor. Spores white, elliptical, smooth, 5–6 × 2–2.5 microns.

Edibility Not known.

Habitat Broadleaf woods, usually under beech in peaty ground.

Season Summer and autumn.

Note The veils which envelop the embryonic carpophores consist of spheroid-shaped cells. Their remains on the cap and stipe have a mealy texture for this reason. Although *C. adulterina* is not found in North America, *C. sistrata* (*Lepiota seminuda*) is a very similar species that is.

25 LEPIOTA CRISTATA

Etymology From Latin, "crested," because of the ornamentation on the cap.

Description Cap 3–4 cm, campanulate then spreading, often with a large obtuse umbo, whitish, fibrillose, with adpressed reddish brown scales that cover the disc consistently. Gills white, free, crowded. Stipe 4–6 × 0.3–0.7 cm, white, fibrillose, tending to turn yellow or pale reddish brown, cylindrical, fragile, almost smooth, with a small membranous ring, often reddish, short-lived. Flesh white, reddish, fine. Strong odor of fruit and acid, flavor unpleasant. Spores white, wedge-shaped, smooth, 6–8 × 3–4 microns, dextrinoid.

Edibility Suspected of having caused cases of poisoning.

Habitat In humus-rich ground, in shady woodland, or in meadows and gardens.

Season From late spring through late autumn.

Note Various small species, with the cap covered in scales or brownish warts, are very poisonous. One such is *L. helveola;* the cap is initially ochre-colored, turning reddish when picked; ring often conspicuous. Grows best in cultivated fields with fodder crops.

---

26 LEPIOTA CLYPEOLARIA

Etymology From Latin, "shield-shaped."

Description Cap 3–8 cm, campanulate then spreading with large obtuse umbo, cuticle brown, breaking up outside disc into large numbers of small scales or granules, margin with festooned velar remains. Gills white or cream-colored, free, soft, and wide. Stipe 6–8 × 0.4–1 cm, covered with cottony down below ring, which is also floccose and not always conspicuous. Flesh white, soft. Slightly fruity odor and sweetish flavor. Spores white, fusiform, smooth, typically 12.5–14 × 4.5–5 microns, dextrinoid.

Edibility Not recommended (see **caution**).

Habitat Woodland.

Season Summer and autumn.

Note This species is recognized by the cottony material on the stipe, and confirmed by an examination of its distinctive, elongated, spindle-shaped spores.

Caution Although edible, this species is one of several small *Lepiotas* about which little is known except that some are poisonous. For this reason, and because of difficulty of identifying it with any certainty in the field, it should *not* be eaten.

27 LEPIOTA ACUTESQUAMOSA

Synonym *Lepiota friesii.*
Etymology From Latin, "sharp-scaled."
Description Cap 8–10 cm, hemispherical then convex, flattened latterly, dark brown covered with downy, pointed, reddish brown warts. Gills white, free, straight, often forked. Stipe 8–11 × 1.5–2 cm, same color as cap, cylindrical or enlarged at the base, squamose at base, with white, descending ring which is pendant. Flesh white tending to turn yellow. Strong somewhat acid odor and fairly unpleasant flavor. Spores white, elliptical, smooth, 7 × 3–4 microns, dextrinoid.
Edibility Not recommended because of the disagreeable flavor.
Habitat In bare ground in gardens, near trees, or in woods of broadleaf trees, in particular oak and beech.
Season Summer and autumn.
Note There are several similar species with raised, pointed cap scales; none is known well enough to recommend as an edible.

---

28 MARASMIUS ANDROSACEUS

Common name Horsehair fungus.
Etymology From a Greek word for an unknown marine plant.
Description Cap 0.4–1 cm, fairly umbilicate, radially striate, whitish or washing to (or entirely) pinkish brown or cocoa-brown. Gills fairly distant, cap-colored. Stipe 3–6 × 0.1 cm, filiform, strong, black, glabrous, straight, twisted and striate when dry, directly inserted in substrate with black rhizomorphs present about base. Flesh virtually nonexistent. No distinctive odor or flavor. Spores white, elliptical, smooth, 6.5–9.5 × 3–4 microns.
Etymology Of no value because of size.
Habitat Grows in large clusters and groups on dead leaves and twigs, and more frequently on pine needles.
Season Spring, summer, and autumn.
Note A similar but foul-smelling species is *M. perforans* (**32**). Another similar species, found primarily in Europe, is *M. splaschnoides,* which has a whitish cap with the disc fairly pinkish yellow and the stipe very villose in the upper part; it grows on dead oak and chestnut leaves. The fruit-bodies of *M. hudsonii* grow on holly leaves, and are entirely covered with long hairs; *M. buxi* is found on box leaves, with short hairs on just the stipe.

29 MARASMIUS FOETIDUS

Synonym *Micromphale foetidum.*
Etymology From Latin, "foul-smelling."
Description Cap 1.5–4 cm, fairly dark brown, membranous, sometimes almost diaphanous, convex becoming flat then de-pressed, radially striate and plicate. Gills reddish, adnate or slightly decurrent, distant and connected by venations. Stipe 1–4 × 0.1–0.2 cm, brown quickly becoming blackish, narrow-ing abruptly toward base, velvety, in rare cases with mycelial formation, roundish at base. Flesh fine, yellowish in cap, blackish brown in stipe. An odor and flavor of putrid water, also slightly garlicky. Spores white, elliptical or pear-shaped, smooth, 8–12 × 4–6 microns.
Edibility Nontoxic, but terrible flavor.
Habitat On rotting branches and twigs in woods.
Season Summer to early winter.
Note *M. inodorus* is very similar, with whitish gills, but has no odor. The nasty odor does, however, occur in some other spe-cies: *M. brassicolens,* on detritus and leaves of beech trees, often in small groups; *M. perforans* (**32**); and *M. acicola,* which forms dense groups of fruit-bodies with the mycelium embrac-ing parts of firs. These species are found primarily in Europe.

30 MARASMIUS RAMEALIS

Synonym *Marasmiellus ramealis.*
Etymology From Latin, "of branches," from its habitat.
Description Cap 0.6–1.5 cm, white, reddish at disc, convex then flat and eventually also depressed, slightly rough. Gills white or pinkish, cream-colored, not very crowded. Stipe 6–10 × 0.1–0.2 cm, whitish, reddish at base, often curved, mealy and squamulose in upper half. Flesh whitish, reddish brown at base. No odor or flavor. Spores white, elliptical, smooth, 8.5–10.5 × 3.4 microns.
Edibility Of no value because of size.
Habitat In groups on dead wood and sticks, in woods.
Season Summer, autumn, and early winter.
Note *M. amadelphus* grows on conifer wood, with a pale ochre cap, whitish at edge, slightly pruinose beneath the mi-croscope, striate; stipe is usually cap-colored. *M. tricolor* and *M. omphaliformis* have a fairly deeply depressed cap and fairly decurrent gills. The former, growing on buried grass stalks, has a white cap, cream-colored gills and the stipe turning from whitish to brown-gray or blackish. The latter species has a brownish yellow diaphanous cap, distant white gills and a yel-lowish stipe. *M. magnisporus* (*M. candidus*) is white and grows in profusion on berry canes.

31 MARASMIUS OREADES

Common name Fairy-ring mushroom.
Etymology From Greek, "of the mountains."
Description Cap 2–6 cm, convex turning eventually flat, sometimes slightly depressed with a slight umbo the color of yellow leather, ochreous, more brightly colored in young specimens, tending to turn pale when drying with striate margin when wet. Gills whitish, free, wide alternating with lamellae. Stipe 4–10 × 0.2–0.4 cm, pale ochre, cylindrical, very sturdy, elastic, solid with mycelium at base. Flesh fine, pale ochre, leathery in stipe. Pleasant cyanic odor, and a tasty flavor. Spores white, elliptical, smooth, 8–10 × 5–6 microns.
Edibility Caps very good, ideal for drying.
Habitat Meadows, grassland, and roadsides, in lines or rings.
Season May through October.
Note Tiny amounts of highly toxic hydrocyanic or prussic acid are sometimes produced when the basidia mature, causing a faint almond odor but not affecting edibility.
Caution Do not confuse with toxic *Clitocybe dealbata* (attached to slightly decurrent gills), nor with whitish species of *Inocybe* (producing a brown spore print).

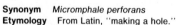

32 MARASMIUS PERFORANS

Synonym *Micromphale perforans*
Etymology From Latin, "making a hole."
Description Cap 0.8–1.2 cm, whitish turning reddish, convex becoming flat and slightly umbonate with striations. Gills whitish, adnate, fairly crowded, connected by venations. Stipe 2–3 × 0.1 cm, brown turning blackish, pinkish at the top, tough and velvety. Flesh virtually nonexistent. Strong odor and flavor of putrid water, with a hint of garlic. Spores white, lanceolate, smooth, 5.5–6 × 3 microns.
Edibility Of no value because of size and flavor.
Habitat On conifer needles, one or two fruit-bodies per needle.
Season Summer and autumn.
Note *Marasmius, Marasmiellus,* and *Micromphale* are three very similar genera that differ microscopically and chemically; there are many species in North America but in very few regions have they been thoroughly collected and studied.

33 MARASMIUS ROTULA

Etymology From Latin, "small wheel."

Description Cap 0.5–1.5 cm, white all over or with central umbilicate part shaded with yellow or gray, membranous, never flat, plicate or with margin regularly crenate. Gills white, adnate, typically attached to a collar which may or may not be attached to stipe. Stipe 2–5 × 0.1 cm, filiform, blackish, horny, striate when dry. Flesh almost nonexistent. No odor or flavor. Spores white, piriform, smooth, 7–10 × 3.5–5 microns.

Edibility Of no value because of size.

Habitat In groups on branches, wood, or dead roots, in woods, heathland, and moorland.

Season Spring to early winter.

Note Almost identical *M. bulliardii* grows on leaves; the center of the small depression is a black or brown point. Two species with the cap umbilicate, grooved and sinuate, and identical gills grow on grass stalks: *M. limosus* and *M. graminum*. The former may grow on *Carex* species (type of sedge), with a brownish ochre cap, uniform and pale, with 6–10-cm gills; the latter has a red, orange-red, or reddish brown cap, often with a brown or blackish papilla.

---

34 COLLYBIA DRYOPHILA

Synonym *Marasmius dryophilus.*

Etymology From Greek, "living among oaks."

Description Cap 2–5 cm, color varying from whitish, ochreous, reddish to brown, often paler toward the edge, diaphanous in wet weather, smooth. Gills quite crowded, white, adnate. Stipe 4–7 × 0.2–0.4 cm, whitish or yellowish or orange-brown, glabrous at the base as well, which often has small mycelial threads, fistulous, tough. Flesh fine, white, ochreous when wet or damp. No special odor or flavor. Spores white, elliptical, smooth, 4.5–6 × 3–4 microns.

Edibility Mediocre.

Habitat In small groups in broadleaf and coniferous woods, also on open ground or rotting wood.

Season Year-round, frequent in summer and autumn.

Note *C. butyracea* is very similar but the cap has a buttery feel and the spore print has a pinkish tinge. In late summer and autumn *C. dryophila* is often found parasitized by what looks like a translucent jelly fungus, but is actually deformed inflated *C. dryophila* tissue covered by a very thin layer of a fungus called *Christiansenia mycetophila*.

35 MARASMIUS ALLIACEUS

Etymology From Latin, "garlicky."

Description Cap 1–4 cm, campanulate, late-spreading and mildly umbonate, eventually striate and grooved radially, turning from whitish to gray-brown with age. Gills whitish, sometimes apparently free, usually adnexed, ventricose, with edge curled when dry. Stipe 4–20 × 0.2–0.4 cm, blackish, tough, rigid, almost cylindrical, pruinous and velvety, extending into rhizomorphs or covered with hairs at base. Flesh white, blackish in stipe. Strong odor and flavor of garlic. Spores white, almond-shaped, smooth, 8–11 × 6–7 microns.

Edibility Caps good for flavoring or seasoning.

Habitat In European beechwoods, on wood or dead leaves, rarely found beneath other broadleaf species.

Season Summer and autumn.

Note The odor is strong and persistent. Other species also have a marked garlicky odor: *M. prasiosmus* with a light brown cap, ochre or brownish stipe narrowing toward top, markedly tomentose, base wooly with mycelium which clings to and embraces the litter, especially with oaks; *M. porreus*, yellowish with striate cap edge, stipe reddish, pubescent, hairy at base, gills yellowish, flesh red and juicy; and *M. scorodonius* (**36**).

---

36 MARASMIUS SCORODONIUS

Etymology From Greek, "garlicky."

Description Cap 1–3 cm, orange-ochre with rapid tendency to lose its color, becoming whitish from the edge, membranous, soon flat, with wavy margin. Gills whitish, adnate, connected by veins, fairly crowded and slender. Stipe 2–5 × 1–0.2 cm, dark red-brown toward base, smooth, glabrous, almost shiny, tough and hard, cylindrical or more often enlarged toward the top. Flesh meager. Strong smell and flavor of garlic. Spores white, lanceolate, smooth, 5–9 × 3.5–5 microns.

Edibility Caps can be used for flavoring.

Habitat On sticks, roots, bits of wood, leaves or needles, in acid soil.

Season Summer and autumn.

Note The caps of the larger species of the *Marasmius* group that smell of garlic are usable, fresh or dried, as flavoring. Some foul-smelling species may have a slight garlicky smell but are inedible, such as *M. perforans* (**32**), *M. foetidus* (**29**) and to a lesser extent *M. impudicus*, which is brown with pinkish shading, more conspicuous on the gills, and has a stipe covered with fine whitish down becoming floccose at the base.

37 COLLYBIA PERONATA

Synonym *Marismius peronatus.*
Etymology From Latin, "booted," from the base shape.
Description Cap 3–7.5 cm, yellowish or pale brick-red eventually light ochre-brown, convex then flat, depressed and umbonate, rough, with a striate cap edge. Gills cream-colored then ochreous brown or cap-colored, adnexed then detached, free, crowded. Stipe 5–9 × 0.2–0.6 cm, cap-colored, narrowing toward top, slightly curved, base covered with evident white or yellowish hairs, solid. Flesh pale yellow, rather leathery. Insignificant odor, but spicy flavor. Spores white, elliptical, smooth, 6–8 × 3–4 microns.
Edibility Used as flavoring because of sharp taste.
Habitat In litter and wood detritus, especially broadleaf.
Season Spring to first cold spells in winter.
Note Carpophores with a pubescent and paler base, yellowish or light nut-brown, belong to a different species, *Marasmius urens*. Neither species is reliably reported in North America. The litter of broadleaf woods plays host to groups of *M. globularis*, with a campanulate cap, liking plenty of moisture; initially white with faint pink or violet shading; stipe leathery, white, supple and brownish at base. It is pleasant to eat, with the same odor as *M. oreades* (31).

---

38 OUDEMANSIELLA RADICATA

Synonym *Collybia radicata.*
Etymology From Latin, "with roots."
Description Cap 3–15 cm, olive-brown, brownish gray, or whitish, campanulate then expanded, umbonate, often irregular, humped, radially rough, very slimy in wet weather, smooth when dry. Gills bright white, with cream-colored or brownish areas at the edge in old specimens, almost free, ventricose, quite distant. Stipe 10–20 cm or more with a diameter of 0.5–1 cm, whitish or pale brown-ochre, narrowing toward top, with an enlarged base extending into the ground with a long rooting appendage, fusiform, smooth at the end, lengthwise striations, cartilaginous. Flesh white, soft in cap, fibrous in stipe. No particular odor or flavor. Spores white, broadly elliptical, smooth, 14–15 × 8–9 microns.
Edibility Mediocre because of cap's sliminess.
Habitat Very common in beechwoods in the litter or rotting wood of tree stumps.
Season Spring to early winter.
Note *O. longipes,* is dry and has a velvety stipe and pruinous, sooty gray-brown cap. *Caulorhiza umbonata* grows under coastal redwoods in northern California, and has a bright orange-brown cap.

39 OUDMANSIELLA MUCIDA

Synonym *Mucidula mucida; Armillaria mucida.*
Etymology From Latin, "viscid" or "slimy."
Description Cap 3–8 cm, pure white or shaded with gray or olive-green at center, thin, almost diaphanous, hemispherical then convex, obtuse, fairly rough radially, with cap edge tend·ing to becoming striate as it becomes thinner. Gills white turning to cream, adnate-decurrent, distant, wide, thin, soft. Stipe 4–7.5 × 0.4–1.5 cm, white with base enlarged and covered with small blackish brown scales, tough, elastic, curved, striate above ring. Ring white, then turning darker when drying, pendent, often grooved. Flesh white, mucilaginous. No particular odor or flavor. Spores white, globose, smooth, 15–17 microns.
Edibility Nontoxic but bad flavor.
Habitat Isolated or in small clusters on beech trunks, more rarely on oak and birch.
Season Summer and autumn.
Note *O. nigra* grows on the ground or on rotting wood in beech stumps, with pruinous cap and stipe, dry, slate-gray. Neither species is found in North America.

40 STROBILURUS TENACELLUS

Synonym *Collybia tenacella.*
Etymology From Latin, "quite tough," because of the consistency of the stipe.
Description Cap 1–3 cm, grayish brown turning paler, very rarely whitish, convex or broadly conical-campanulate, then flat, sometimes slightly raised at center, moist. Gills grayish then white, with frequent tendency to turn yellow, adnexed, crowded. Stipe 2–6 × 0.2–0.3 cm, white, soon turning ochre-brown, white at top, cylindrical, with base extending into a dark brown fibrillose mycelial thread, downy. Flesh leathery and tough in stipe, softer in cap, white. Slightly mealy odor, somewhat bitter flavor. Spores white, elliptical, smooth, 4–5 × 2.5–3 microns.
Edibility Mediocre.
Habitat On buried or moss-covered pinecones or on spruce cones.
Season Late winter, early spring.
Note This is a very early common species in Europe. Although not found in North America, several macroscopically identical species do occur here, including *S. trullisatus*, common in the Northwest on buried cones of Douglas fir; and *S. conigenoides*, common in the Southeast on fallen magnolia fruits.

41 COLLYBIA CONFLUENS

Synonym *Marasmius confluens.*

Etymology From Latin, "confluent," because of the tuftlike stipes.

Description Cap 2–4 cm, campanulate, convex then flat, slightly raised at center, yellowish white, with reddish or pinkish shades, slightly hygrophanous, silken when dry, cuticle not detachable, cap edge often wavy when old. Gills very crowded, small, white turning to cream, pale nut-brown. Stipe 6–12 × 0.2–0.3 cm, cylindrical, often slightly curved, same color as cap, downy from top to bottom, hollow, compressed, flared and pruinous at top, leathery. Flesh whitish, fine. Odor initially cyanic, then slightly garlicky. Spores white, elliptical, smooth, 7–12 × 3–6 microns.

Edibility Only the caps are good.

Habitat In clusters in broadleaf woods.

Season Summer and autumn.

Note *C. hariolorum,* if a distinct species, forms tuftlike fruitbodies on the leaf litter, especially beneath beech. The stipe is finely velvety, downy at base, whitish or cream-colored, often spotted with red. Gills crowded but longer than those of *C. confluens.* The odor is garlicky or, more precisely, like camembert cheese.

42 COLLYBIA CIRRHATA

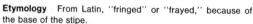

Etymology From Latin, "fringed" or "frayed," because of the base of the stipe.

Description Cap 0.5–1 cm, white, reddish brown at center, or ochreous, initially convex then flat, slightly depressed, often with a very small central protuberance, cuticle slightly silken, often broken up concentrically when mature, slightly grooved. Gills white, crowded, unequal, adnexed, separate from stipe when mature. Stipe 2.5–5 × 0.1 cm, whitish, rather filiform, supple with powdery surface that extends to the base in a tuft of interwoven fibrils. Flesh white, very meager. Spores white, elliptical, smooth, 4–5 × 2 microns.

Edibility Of no value because of size.

Habitat On the remains of old mushrooms, in woods among dead leaves.

Season Summer and autumn.

Note *C. cookei* and *C. tuberosa* are very similar and found in the same habitat; the former originates from a roundish, yellowish orange sclerotium, the latter from an appleseed-shaped, brownish sclerotium.

43 COLLYBIA BUTYRACEA

Etymology From Latin, "of butter" or "buttery."
Description Cap 5–8 cm, reddish brown, purplish gray, double-layered, tending to turn yellow when dry, convex then expanding, fairly umbonate, smooth, greasy to the touch. Gills white, slightly adnexed, sometimes free, thin, crowded, with a crenate edge. Stipe 5–8 × 0.5–1 cm, ochreous or grayish, conical and narrowing toward top, white and downy at the base, with a rigid cuticle, cartilaginous, striate, smooth, rarely downy. Flesh pinkish or very light brown, then white, soft, and watery. Rancid odor, sweetish flavor. Spores whitish, creamy pink in mass, arched, smooth, 6.5–8 × 3–3.5 microns, slightly dextrinoid.
Edibility Mediocre.
Habitat In woods, with a preference for conifers.
Season Summer and autumn.
Note *C. dryophila* is very similar but lacks a buttery feel to the cap, and its spores are white in mass.

44 COLLYBIA FUSIPES

Etymology From Latin, "spindle-footed."
Description Cap 4–10 cm, reddish brown tending to fade or darken, convex then flat, center slightly raised, smooth, dry, edge often incised. Gills whitish, becoming cap-colored with dark brown spots, adnexed then separate and almost free, distant, connected by veins, curled. Stipe 7–15 × 1 cm, cap-colored, cartilaginous, enlarged, ventricose at center, narrowing at top and bottom, pointed at base, compressed, often curved, striate-grooved lengthwise, almost blackish at base. Flesh cap-colored, fading with age, firm. Faintly rancid odor, sweet flavor. Spores whitish, creamy pink in mass, elliptical, smooth, 5–6 × 3–4 microns, slightly dextrinoid.
Edibility Mediocre, see **note** below.
Habitat Tufflike at base of oaks or chestnuts, or old tree stumps.
Season Summer and autumn.
Note The stipes of the carpophore grow from the remains of the fruit-bodies of previous years, which form a sort of hypogeal sclerotium. This species can sometimes cause stomach upsets, not least because the fruit-bodies which are dead remain for long periods virtually unaltered on the ground. This species is not reliably known to occur in North America.

45 COLLYBIA MACULATA

Etymology From Latin, "spotted."
Description Cap 7–12 cm, white then with reddish spots, sometimes becoming reddish all over, fleshy, compact, flat-convex, obtuse, smooth, with slender cap edge, initially invo-lute. Gills cream-colored, often with reddish spots, semifree, straight, very crowded, denticulate. Stipe 7–12 × 1–2 cm, with reddish spots, sometimes ventricose, narrowing toward the base, compact, extremely fibrous, striate, cartilaginous, solid. Flesh firm and white. Flavor somewhat bitter or unpleasant. Spores whittish, creamy pink in mass, subglobose, pointed, smooth, 5–6 microns, slightly dextrinoid.
Edibility Mediocre, but must be boiled.
Habitat In woodland, often gregarious.
Season Summer and autumn.
Note There are several varieties of this species: *scorzon-erea,* with gills and stipe turning yellow; *immaculata,* which does not become spotted; *immutabilis,* which has white gills.

46 FLAMMULINA VELUTIPES

Synonym *Collybia velutipes.*
Etymology From Latin, "velvet-footed."
Description Cap 2–10 cm, orange-red, yellower at edges and dark at center, convex then flat, often eccentric, irregular, lobate, cuticle smooth, viscid, with slightly striate margin. Gills pale yellow tending to light nut-brown, slightly adnexed, un-equal. Stipe 5–10 × 0.4–0.8 cm, cylindrical and slightly ta-pered toward base, curved or twisted, yellow, velvety all over, soon becoming blackish brown or black starting from the base, cartilaginous, tough. Flesh pale yellow, fine, watery and soft in the cap, leathery and fibrous in the stipe. Odor and flavor insig-nificant. Spores white, elliptical, smooth, 7–10 × 3–5 microns.
Edibility Caps quite tasty.
Habitat On dead wood, mainly stumps and roots or in wounds of living trees, often tuftlike.
Season Late autumn to early spring, rarely in summer except in mountains.
Note One of the few epigeal fungi found in winter in temper-ate regions. Easy to cultivate and is marketed in oriental food stores.

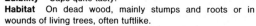

47 COLLYBIA RACEMOSA

Etymology From Latin, "branched."
Description Cap 0.5–1.5 cm, grayish, thin, convex then flat, often imperfectly formed, with a papilla at the disc, downy, striate. Gills grayish, adnate, small and quite crowded. Stipe 3–5 × 0.1 cm, gray with numerous single ramifications, almost perpendicular to the main axis, terminating in a capitate spheroid formation, transparent, slightly glutinous, producing asexual spores. The stipe grows from a black sclerotium buried in the substratum. Flesh grayish. Odor insignificant. Spores white, elliptical, smooth, 5 × 2.5–3 microns.
Edibility Of no value because of size.
Habitat Grows on old mushrooms, especially large species of *Russula* and *Lactarius*.
Season Late summer and autumn.
Note This is a small, rather rare fungus with features that make it easy to identify. It may also live like a simple saprophyte with host carpophores.

48 MELANOLEUCA CNISTA

Synonym *Tricoloma cnista; Melanoleuca evenosa.*
Etymology From Greek, "incised," because the skin of the cap is sometimes cracked.
Description Cap 6–15 cm, whitish gray then cream-white and darker at center, convex then flat around umbo, fleshy, cuticle easily detachable, shiny when dry, tough, with fairly concentric crack lines. Gills whitish cream-color, crowded, unequal, separable, uncinate, not venose. Stipe 6–10 × 0.6–1.5 cm, color of gills, pruinous at top, solid, hard, quite slender, subbulbous at base, fibrillose. Flesh white turning to brownish when exposed to air, tender, fibrous in stipe, thick at center. Odor of aniseed, flavor first sweet then bitter. Spores white, elliptical-ovoid, verrucose, 8–10 × 4–4.5 microns, amyloid.
Edibility Good, but not to everybody's taste.
Habitat Gregarious, in rings or lines, in wet riverside meadows, and in mountain pastures. Northern distribution.
Season An early species, spring to autumn.
Note In meadows large carpophores of *M. grammopodia* grow, with a brownish rounded cap and fibrillose striate stipe. *M. subbreviges* is possibly a variety of it, but has a cap up to 30 cm in diameter, with a short fragile stipe of 5–7 × 1–1.3 cm. Although eminently edible, their strong odor puts some people off.

49 MELANOLEUCA MELALEUCA

Synonym *Tricholoma melaleucum.*
Etymology From Greek, "black and white," because of the contrast between gills and cap.
Description Cap 4–10 cm, gray-brown or soot-colored, brighter in wet or damp weather, convex then flattened, with fairly conspicuous central umbo, fleshy. Gills snowy white, crowded, ventricose, worn at edges. Stipe 5–10 × 0.5–1.2 cm, cap-colored, covered with brownish fibrils, base darker and slightly bulbous, covered with whitish fluff, solid, elastic. Flesh whitish in dry weather, pale ash-gray when wet, yellowish in stipe, turning brownish and rather leathery with age. No particular odor or flavor. Spores white, elliptical, spotted, 7–10 × 5–6 microns, amyloid.
Edibility Fair.
Habitat Grows in clearings and meadows, and in both broadleaf and coniferous woods.
Season Summer and autumn.
Note There are several similar species but few are well known in North America.

50 ASTEROPHORA LYCOPERDOIDES

Synonym *Nyctalis asterophora.*
Etymology From Greek, "like a puffball," from the odor.
Description Cap 1–2 cm, first white or whitish then quickly turning brownish ochre, becoming cottony and powdery after the formation of asexual spores (clamydospores). Gills white, then brown, adnate, distant, thick, sometimes forked or absent. Stipe 1–3 × 0.3–0.8 cm, white then grayish brown, pruinous, solid. Flesh dark grey. Odor and flavor insignificant. Sexual spores white, ovoid, smooth, 5–6 × 3.5–4 microns. Asexual spores brown, spiny, winged, roundish, 12–18 microns.
Edibility Of no value because of size.
Habitat Isolated or in groups, with a preference for the large carpophores of blackening *Russulas* (in the *Russula nigricans* group).
Season Summer and autumn.
Note A similar fungus is *A. parasitica*, which also grows on *Russulas* of the *R. delica* group, or on white *Lactarius* species. It is more tapered and has a gray, silken cap with radial fibrils; and never becomes powdery. It also produces clamydospores which are fusiform and smooth in the gills among the basidia.

51 CYPTOTRAMA CHRYSOPEPLUM

Synonym *Collybia lacunosa.*
Etymology From Greek, "richly gilded."
Description Cap 1.5–3 cm, fairly fleshy, subglobose, convex then expanded, margin slightly striate, dry, orange-yellow, covered with spines which initially form pyramidal warts joined at the apex, then separate to form a hairy layer of the same color. Gills whitish then pale yellow, quite distant, adnate-decurrent. Stipe 2.5–5 × 0.2–0.4 cm, sometimes a bit eccentric, golden or sulfur-yellow, almost cylindrical, solid, surface squamulose, then furfuraceous. Flesh yellowish white, orangish at base and beneath cuticle. No distinctive odor or flavor. Spores white, elliptical, smooth, 10–12 × 3.5–4.5 microns.
Edibility Uncertain.
Habitat Isolated or in small groups on rotting wood in tropical forests, and in eastern North America north into Canada.
Season In rainy periods.
Note The early stages of the carpophores resemble those of the small and conspicuously aculeate species of *Lycoperdon* (puffballs). Quite common, and easily recognizable in its habitat, this fungus has been included in many different genera, and has had many names.

---

52 CATATHELASMA IMPERIALE

Synonym *Biannularia imperialis; Armillaria imperialis.*
Etymology From Latin, "imperial," from its size.
Description Cap 5–20 cm, hemispherical to flat, often with a wavy cap edge, cuticle dry, brown, at first having cottony velar remains, tending to crack from the center outward, margin involute. Gills crowded, small, decurrent, whitish to light hazel, with blackish edge. Stipe 6–12 × 2–5 cm, squat, solid, pointed at the base, whitish or ochreous, with two membranous rings, the upper deriving from the partial veil, the lower from the universal veil. Flesh firm, hard, and white. Strong mealy and watermelonlike odor, flavor pleasant but slightly astringent. Spores white, elliptical, smooth, 12–15 × 5–7 microns.
Edibility Mediocre; good when preserved in vinegar or oil.
Habitat Beneath conifers, in groups, deep in the ground.
Season Summer and autumn.
Note *C. ventricosum* is more common and more widely distributed across northern North America; it is white to grayish, is found beneath spruce, and is considered to be the more tasty of the two species.

53 LYOPHYLUM DECASTES

Synonyms *Tricholoma aggregatum; Lyophyllum aggregatum; Clitocybe multiceps.*
Description Cap 5–15 cm, grayish or leather-colored, fleshy, convex then expanded, often depressed and wavy, smooth. Gills whitish, crowded, adnate. Stipe 4–7 × 0.5–1.5 cm, cylindrical, tough, whitish, pruinose at top, solid. Flesh whitish, firm, fairly leathery in stipe. Slightly mealy odor, sweetish flavor. Spores white, roundish, smooth, 5–7 × 5–6 microns.

Edibility Very good, ideal for preserving in liquid.
Habitat In tuftlike groups, rarely isolated, in woods and open places near trees, also on tree stumps and piles of sawdust and along roadside.
Season Spring to late autumn.
Note The name *decastes* embraces various different fungi which can be considered as different species. *L. loricatum* has a cap which is initially blackish, then the cuticle breaks up into tiny granules as the surface expands, revealing the white background and giving the fungus a spotted or speckled look. *L. fumosum* has a cap which fades and the gills tend to become grayish. The stipes can be found growing together or ramified. Poisonous Entoloma species have salmon-pink prints.

---

54 TRICHOLOMA BATSCHII

Etymology After the German mycologist Batsch.
Description Cap 6–9 cm, chestnut-brown, reddish, convex becoming flattened, initially viscous, then faintly fibrillose when dry, with smooth margin. Gills adnate, whitish then developing dense reddish spots. Stipe 5–10 × 1–2 cm, solid, cylindrical or slightly narrowing toward top, white and furfuraceous uppermost, the rest cap-colored. Flesh white, spotted reddish at edge or where eaten by larvae. Odor of watermelon or bugs, very bitter flavor. Spores white, spheroid, smooth, 5–6 microns.

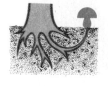

Edibility Inedible because of flavor.
Habitat In broadleaf woods, especially oak, chestnut.
Season Summer and autumn.
Note The clearly two-colored stipe delimits whitish top in the form of a ring. Similar species include: *T. aurantium*, orange-red with a sometimes pungent green-corn odor; *T. albobrunneum*, reddish brown with conspicuous zonation, a dry, fibrillose cap, somewhat bitter, beneath conifers in mountainous areas; *T. pessundatum*, smelling of flour, beneath conifers; *T. ustaloides* (**66**); *T. ustale*, beneath broadleaf trees, bitter. The last three species have a viscous cap and inconspicuous zonations.

55   TRICHOLOMA COLUMBETTA

Etymology   From its former name in France.

Description   Cap 6–10 cm, snowy white, first ovoid or campanulate then flat, sometimes a central umbo, fleshy, slightly viscous, shiny, silken, cuticle detachable, margin at first curled toward the gills then unfolding and becoming undulate, sometimes splitting in the direction of the gills. Gills white, crowded, unequal, fairly small, adnate. Stipe 4–8 × 0.8–2 cm, white with pale bluish green areas or more rarely reddish at the base where it is often more slender, more rarely with a squat base, solid, fibrous and sericeous. Flesh very white, soft, fibrous in stipe, unchanging. Slight odor, flavor sweetish. Spores white, elliptical, smooth, 5–7 × 4–5 microns.

Edibility   Good but slightly fibrous.

Habitat   In groups in woods on fine soil, beneath broadleaf species.

Season   Late summer to late autumn.

Note   The absence of strong odor and sharp flavor should prevent confusion with *T. album*, also lacking the typical greenish blue marking at base of stipe. *T. resplendens* is also white, although cap has fairly bright shades of ochre, with no markings at base, and the odor is pleasant and aromatic.

56   TRICHOLOMA ALBUM

Etymology   From Latin, meaning "white."

Description   Cap 5–12 cm, white, sometimes ochreous at the disc, convex turning flat, fleshy, glabrous, with the margin initially incurved. Gills white, not very crowded, adnate. Stipe 7–8 x 1–1.5 cm, sometimes very long, narrowing toward the top, at times sinuate and also narrowing toward the base when growing deep in the leaf litter, mealy at the tip, rarely striate, solid. Flesh white, soft and fragile in the cap, fibrous in the stipe, with a strong smell of gas and flour when broken. Flavor at first just a little bitter, then hot. Spores white, elliptical, smooth, 4.5–7 x 3–4.5 microns.

Edibility   Inedible, probably slightly poisonous.

Habitat   Broadleaf or mixed woods.

Season   Summer and autumn.

Note   *T. inamoenum* grows underneath conifers in mountainous areas; it also has a strong smell of coal gas. Stipe and gills, which are fairly distant, are both white, whereas the cap has a deeper ochreous color at the disc. Other species with a strong smell—either mealy or rancid—include *T. sulphureum* and *T. lascivum,* which are described in detail, and *T. bufonium.*

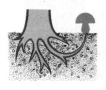

57 TRICHOLOMA FLAVOVIRENS

Synonym *Tricholoma equestre.*
Etymology From Latin, "greenish yellow."
Description Cap 5–10 cm, sulfur- or olive-yellow, reddish at the center due to the tiny scales covering it, convex then turning to flattened and undulate, fairly raised umbo, cuticle detachable, fleshy, smooth, slightly viscid especially when immature, margin thin and curled. Gills lemon-yellow or sulfur-yellow, crowded, of average size, unequal, worn near stipe to which they adhere slightly if at all, edge undulate. Stipe 3–8 × 0.6–1.8 cm, sulfur-yellow with olive-colored areas, sturdy, squat at the base, solid, fibrillose. Flesh whitish, pale yellow beneath the cuticle, thick at the center, thin at the edge, soft. Odorless, with a mealy flavor. Spores white, elliptical, smooth, 6–7 × 4–5 microns.
Edibility Very good, and much sought after.
Habitat In lowland and mountains, especially under conifers, more rarely under broadleaf species, prefers sandy soil.
Season Late summer and autumn.
Note In addition to the above, several varieties are found in North America; *leucophyllum,* with white gills, and *albipes,* with a white stalk.

58 TRICHOLOMA CALIGATUM

Synonym *Armillaria caligata.*
Etymology From Latin, "with the foot covered."
Description Cap 11–20 cm, first hemispherical with an involute edge, then convex, obtusely umbonate, gray-, ochreous, or reddish brown, with radial coloration, smooth with markedly adpressed fibrous scales on a white background, viscous in wet weather, shiny when dry. Gills adnate-rounded, cream-colored, with reddish brown spots when older, large, and crowded. Stipe 9–20 × 2–3 cm, cylindrical or tapered toward bottom, white above membranous ascending ring, below it cap-colored with age, with scaly areas resembling girdles. Flesh white, soft, fairly fibrous in stipe. Pleasant, slightly fruity odor, sweetish flavor with a somewhat bitter aftertaste. Spores white, oval, smooth, 5.5–8 × 4–5.5 microns.
Edibility Excellent.
Habitat Gregarious in coniferous woods, in cold temperate climate.
Season Summer and autumn.
Note This species is very common in the Northern Hemisphere, and several varieties are found in North America; some have no odor, and some are quite unpalatable.

59   TRICHOLOMA IMBRICATUM

Etymology   From Latin, "like roof tiles," from scaly cap.
Description   Cap 4–10 cm, reddish brown, convex then expanded, obtusely umbonate, smooth and fleshy at center, cuticle dry, scaly, cracked, edge involute then thin and wooly. Gills white with reddish spots, quite distant, large, worn near stipe. Stipe 5–10 × 1–1.5 cm, white and pruinous at top, whitish with brown at the base, cylindrical but quickly becoming hollow, fibrillose. Flesh white, brownish at base of stipe, thick only at the center. Odorless, sometimes slightly bitter aftertaste. Spores white, oblong, smooth 5–7 × 4–5 microns.

Edibility   Mediocre, only a few specimens suitable.
Habitat   In small groups in pinewoods, also beneath larch, in lowland and mountains.
Season   Late autumn.
Note   Other species are common in coniferous woods and have distinctive features: *T. aurantium* has an orange-red cap and almost the entire stipe is covered with small scales of the same color, which form a band at the glabrous top, and its flesh is bitter and reddens in the stipe; *T. psammopum*, frequent beneath larch, has an ochreous-red cap, downy with tufted fibrils. The stipe is covered with ochreous or orange cottony scales.

---

60   TRICHOLOMA LASCIVUM

Etymology   From Latin, "licentious," from its odor.
Description   Cap 5–9 cm, light ochreous brown at the disc, fading to white at the edge, fleshy then flat, eventually slightly depressed, delicately pubescent then smooth, dry, with the margin initially involute. Gills whitish, curved and adnexed, eventually adnexed-decurrent, not always straight, crowded. Stipe 7.5–11 × 1 cm, whitish, turning slightly darker when rubbed or with age, rigid, fibrillose, with the top white and pruinous, downy at the base. Flesh thick, firm and white. Very pervasive odor of lilacs, flavor somewhat bitter. Spores white, elliptical, smooth, 6–7 × 3.5–5 microns.

Edibility   Can be eaten.
Habitat   In the litter of beech and oak woods.
Season   Summer and autumn.
Note   *T. inamoenum*, another member of this genus, emits a very strong but not unpleasant smell of acetylene. It grows in coniferous woods in mountains, with a whitish cap with yellow shading. Neither is known in North America.

61   TRICHOLOMA PORTENTOSUM

Etymology   From Latin, "portentous."
Description   Cap 4–12 cm, dark gray with olive-yellow color-ation, fine radial striations throughout changing from violet to black, convex then flat and umbonate, edge thin, often cracked radially, cuticle sticky and easily detachable. Gills initially whit-ish then pale yellow, distant, unequal, quite large, worn near stipe. Stipe 5–10 × 1–2 cm, whitish with pale yellow or green-ish gray shades, cylindrical, enlarged at base, first solid then hollow, fibrillose. Flesh pale yellow or white, grayish beneath cuticle, thick at center, thin at edge, fragile. Slight odor and flavor of flour. Spores white, elliptical, smooth, 6–8 × 3–4.5 microns.
Edibility   Very good.
Habitat   Grows alone or in groups in coniferous and broad-leaf woods.
Season   Late, from autumn to the first frosts.
Caution   Some gray-capped species of *Tricholoma* are poi-sonous, so identify with great care.

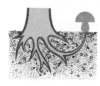

62   TRICHOLOMA SAPONACEUM

Etymology   From Latin, "soapy," because of the odor.
Description   Cap 5–10 cm, gray-brown or greenish or of vari-ous other coloration, convex becoming flat, with a tendency to crack in dry weather, slightly greasy or viscous if wet, with curled margin. Gills white shading to pale blue, with a tendency to develop reddish spots, distant and adnate. Stipe 5–10 × 1.5–2 cm, whitish, reddish from base, often curved, narrowing at the base, rooting. Flesh white, sometimes reddening particu-larly at the base, soft. Odor sometimes of lavender or soap, flavor sweetish. Spores white, elliptical, smooth, 5–6 × 3–4 microns.
Edibility   Can be eaten, but not very pleasant.
Habitat   Gregarious, sometimes almost tuftlike in coniferous and broadleaf woods.
Season   Summer and autumn.
Note   A very variable species, *T. saponaceum* is always iden-tifiable by the reddish area at the base of the stipe.

63 TRICHOLOMA SEJUNCTUM

Etymology From Latin, "separate."
Description Cap 7.5–10 cm, pale greenish yellow with blackish fibrils arranged radially, darker at disc, pale yellow or whitish toward margin which is curled or involute, turning from convex to expanded, humped, rarely umbonate, slightly viscid in damp weather. Gills white, often with yellowish shading, quite distant, adnate. Stipe 6–12.5 × 1.5–2.5 cm, white with yellow markings, ventricose then almost cylindrical, minutely furfuraceous at top. Flesh white, pale yellow beneath cuticle and on sides of stipe, fragile. Mealy odor, slightly bitter flavor. Spores white, subglobose, smooth, 5–7 microns.
Edibility Edible but not tasty.
Habitat In broadleaf woods, especially with oak and chestnut, and in mixed woodland.
Season Summer and autumn.
Note The cap is the same color as *Amanita phalloides* (4), but speckled or streaked toward the edge. An ochre-brown species, *T. coryphaeum* has a yellowish cap with brownish squamulae, gills white with edge turning yellow and stipe with yellow marking in the central area. It has the same smell and flavor as *T. sejunctum*, and prefers beechwoods.

64 TRICHOLOMA TERREUM

Etymology From Latin, "earth-colored."
Description Cap 4–8 cm, campanulate or conical, then flat, fairly umbonate, often cracked, covered with small silken radial scales, grayish or brownish, cuticle detachable, tough. Gills whitish then pale ash-colored, fairly distant, long and large, unequal, worn near stipe, fragile, margin denticulate. Stipe 3–5 × 0.5–1 cm, cylindrical, solid then subfistulous, fragile, white shaded with pale ash-gray, grayish veil fairly conspicuous, quite long-lasting. Flesh grayish-white, thin, fragile. Mild odor, sweetish flavor, somewhat bitter when mature or old. Spores white, elliptical, smooth, 5–8 × 4–5 microns.
Edibility Very good, particularly when young.
Habitat In groups, almost tuftlike, or in rings in woodland, especially beneath conifers.
Season Summer to early winter.
Note This is a complex of species, not all of which are edible, having a dry grayish cap with radial scales or fibrils. *T. triste* has odorless flesh and a large partial veil shaped like a curtain which leaves no traces on the stipe but remains as tufts hanging from the cap edge.

65 TRICHOLOMA PARDINUM

Synonym *Tricholoma tigrinum.*
Etymology From Latin, "striped."
Description Cap 6–8 cm, slate- or silver-gray or dirty white, slightly suffused with violet, sometimes darker at center, hemispherical or campanulate then convex-obtuse or slightly umbonate, sometimes depressed when old, fleshy, cuticle easily detachable with silky brownish scales, adpressed, rectangular, denser at the center, fairly concentric, margin curled then straight, often cracked or split. Gills yellowish white with olive-green shading, crowded, unequal, long, worn near stipe, to which they are only slightly adherent, if at all. Stipe 4–10 × 1–4 cm, white and pruinose at top, fibrils halfway down or covered with small brownish scales, base often enlarged and fairly dark brownish, sturdy, solid, and fibrous. Flesh grayish-white in cap, yellowish at base of stipe. Faint mealy odor and flavor. Spores white, ovoid, smooth, 8–10 × 6–7 microns.
Edibility Poisonous.
Habitat Under northern conifers.
Season July through October.
Note Its close resemblance to *T. terreum* (**64**) makes this a particularly dangerous mushroom. It causes severe gastric upset.

66 TRICHOLOMA USTALOIDES

Etymology Similar to *T. ustale,* from its "burnt" color.
Description Cap 5–10 cm, thick, fleshy, compact almost hard, almost hemispherical becoming eventually flat, slightly raised at disc but more often depressed, with incurved margin, quite often ribbed at edge, bright reddish brown, not fading, shiny, viscid then dry, with innate scales resembling stains. Gills white with reddish highlights, and reddish spots when mature, margin irregular, adnate, not very crowded. Stipe 6.5–10 × 1–1.5 cm, solid, slightly cavernous in upper part when mature, cylindrical or enlarged or fusiform at base, reddish brown or reddish ochre, darker toward base which is shaded with white at top, pruinose, otherwise with darker squamulae and fibrils. Flesh white, tending to redden especially where eaten by larvae, thick and solid. Strong mealy odor and flavor. Spores white, subglobose or broadly elliptical, smooth, 6–7 × 4.2–5 microns.
Edibility Fair.
Habitat Common in broadleaf woods.
Season Late summer and autumn.
Note *T. ustale,* also edible, lacks a mealy smell, and has a chestnut-brown cap, ochreous brown at the edge; its stipe is white with reddish fibrils when mature. Neither is reliably known to occur in North America.

67 TRICHOLOMA VIRGATUM

Etymology From Latin, "striped" or "barred" because of the fibrils on the cap.

Description Cap 4–8 cm, ash-gray with violet shading, darker at the center, conical then expanded, fairly pointed umbo, cuticle detachable with dark, silky fibrils. Gills grayish white, fairly crowded, uncinate. Stipe 5–10 × 0.5–1.8 cm, whitish, cylindrical, smooth and solid. Flesh whitish. Odorless, with a pungent flavor after prolonged chewing. Spores white, elliptical, smooth, 6–7 × 4–5 microns.

Edibility This species is suspected of causing mild cases of poisoning.

Habitat Grows in particular under fir and beech.

Season Summer and autumn.

Note The flesh has a particular sharp taste, which nevertheless differs from that of other fungi like the *Russula* and *Lactarius* groups; the effect resembles being pricked with tiny needles at the tip of the tongue.

---

68 TRICHOLOMA SULPHUREUM

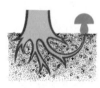

Etymology From Latin, "sulfur-colored."

Description Cap 3–9 cm, sulfur-yellow, often brownish at the center, convex then flat, sometimes with a slight central umbo, sometimes slightly depressed, silky, dry, glabrous at the edge, which is involute. Gills sulfur-yellow, quite distant, long and large, worn near the stipe. Stipe 2–8 × 1–2 cm, sulfur-yellow, enlarged toward the base, striate lengthwise, solid at first becoming hollow, fibrous. Flesh sulfur-yellow, sometimes lighter, fibrous. Odor of acetylene or coal gas, flavor mild. Spores white, elliptical, smooth, 8–10 × 5–6 microns.

Edibility Suspected of having caused cases of mild poisoning.

Habitat In broadleaf and coniferous woods.

Season Late summer to late autumn.

Note In rare cases the gills may be pinkish or lilac-colored, typically in the variety *rhodophylium*. *T. bufonium*, which is very similar, can be considered a variety of *T. sulphureum*, with a brown or reddish brown cap. The strong smell distinguishes it from *T. flavovirens* (**57**), one of the most sought-after species.

69 CALOCYBE IONIDES

Synonym *Tricholoma ionides; Lyophyllum ionides.*
Etymology From Greek, "violetlike."
Description Cap 3–6 cm, deep violet, often brownish blue at disc, fading with age, fleshy, convex-campanulate then flat, often obtusely umbonate, initially pruinous soon becoming smooth except at incurved cap edge. Gills white tending to turn pale yellow, adnate, crowded, thin. Stipe 3–6 × 0.7–1.2 cm, cap-colored or paler, tough, elastic, narrowing or slightly enlarged at base, surface fibrillose or striate and fibrous, mycelium white, cottony or hairy at base, solid. Flesh white, violet beneath cuticle and at base of stipe. Mealy odor and flavor, at times strong. Spores white, elliptical, smooth, 5–6.5 × 2–3 microns.
Edibility Good.
Habitat Gregarious, in leaf litter beneath beech and oak, also found in open grassland and under conifers.
Season Summer and autumn.
Note In rare cases the cap may be pinkish or cream-colored. Especially beneath conifers, seldom in broadleaf humus, is *C. obscurissima,* a more slender species with a violet-brown cap and stipe. Neither is known in North America.

70 CALOCYBE CARNEA

Synonym *Tricholoma carneum.*
Etymology From Latin, "flesh-colored."
Description Cap 2–3 cm, reddish then pale flesh-colored, paling further, shiny, thin, hemispherical becoming flat, rarely raised at the center, smooth and dry. Gills pure white, rounded near the stipe, virtually free, very crowded. Stipe 2–5 × 0.2–0.6 cm, same color as cap, narrowing toward the base, tough, almost cartilaginous, sometimes pruinous at the top. Flesh white, thin. Odorless, no particular flavor. Spores white, elliptical, smooth, 4–6 × 2–3 microns.
Edibility Mediocre.
Habitat In meadows, and fields, rarely in woods.
Season Summer and autumn.
Note *C. cerina* is shiny in appearance, with a beeswax color or reddish brown, gills yellow and stipe yellowish, darkening at the base. *C. persicolor* has a distinctive pink coloration, like peaches; it grows in tufts in well-manured fields or by the roadside. The cap and stipe are the same color but the gills are white. *C. alpestris* is another pinkish species, but has cream-colored gills.

71 CALOCYBE GAMBOSA

Synonym *Tricholoma georgii.*
Common name St. George's mushroom.
Etymology From Latin, "large-legged."
Description Cap 5–10 cm or more, hemispherical then convex, very fleshy and thick, then distended, normally humped, dry, finely velvety, color from white to buff to yellowish, margin smooth, pruinous, curled, only later expanded and often sinuate. Gills crowded, thin, whitish then cream-colored, very small at first, adnate. Stipe 3–6 × 1–2 cm or more, sturdy, cylindrical but more often enlarged at base, whitish with ochreous markings toward bottom, fibrillose, pruinous at top, solid. Flesh compact, thick, white. Strong, pleasant smell of fresh meal, similar flavor but with resinous aftertaste. Spores white, elliptical, smooth, 5.5–6 × 3.5–4 microns.
Edibility Excellent and highly prized.
Habitat In grassland in hilly or mountainous areas, prefers calcareous soils, in groups, lines, or rings, often beneath brambles, also in broadleaf woods.
Season Spring.
Note Not known in North America.

72 TRICHOLOMOPSIS RUTILANS

Etymology From Latin, "reddening."
Description Cap 5–20 cm, yellowish, thickly covered with reddish granulations which turn into small scales, fleshy, campanulate then convex and flat, often umbonate with involute edge, slightly grooved. Gills sulfur-yellow, adnate, crowded and broad. Stipe 4–14 × 1–2.5 cm, uniform or enlarged or tapered at base, yellow, fairly covered with small reddish scales, pruinous at top. Flesh thick, soft, leathery in stipe, yellow. No particular odor or flavor. Spores whitish, oval or spheroid, smooth, 5–8 × 4.5–6 microns.
Edibility Can be eaten, but taste unpleasant.
Habitat Isolated or in groups on tree stumps or roots of conifers.
Season Late spring to late autumn.
Note Other less common and smaller species grow on wood. *T. flammula* has a sulfur-yellow cap covered with tiny violet-red scales. *T. decora,* whose cap can reach 12 cm in diameter, is golden yellow with blackish brown scales, at least at the center. *Oudemansiella platyphylla,* with a cap striped with grayish brown or blackish innate fibrils, very long, large gills, and a large pithy stipe with rhizomorphs at the base, is included by some in the genus *Tricholomopsis.*

73 CLITOCYBE GLAUCOCANA

Synonym *Tricholoma glaucocana; Lepista glaucocana; Rhodopaxillus glaucocana.*

Etymology From Latin, "whitish, pale blue-green."

Description Cap 6–12 cm, convex-campanulate or flat and slightly umbonate, fleshy, blue-gray, lilac, violet, all pale, tending to fade more with age, disc shaded ochreous, margin involute. Gills cap-colored, quite crowded, erose and adnate, detachable from cap. Stipe 6–10 × 1–2 cm, cap-colored, pruinous at top, otherwise somewhat fibrillose, with an enveloping feltlike mycelium at the enlarged base, solid. Flesh whitish with pale lilac, violet, pale blue-gray shading, thick. Odor often loamy with an acrid-sweetish flavor. Spores whitish, shaded lilac in mass, elliptical, minutely roughened, 6–8 × 3–5 microns.

Edibility Good.

Habitat Gregarious, on leaf litter in woods.

Season Summer and autumn.

Note *C. saeva* grows in open places and in grassland, during late autumn, often in lines or rings; it has a pale brownish ochre cap, gills shaded with pale lilac and the stipe violet-lilac. *C. irina,* which is creamy flesh-colored all over, has a gentle smell of orange or iris flowers.

74 CLITOCYBE NUDA

Synonym *Tricholoma nudum; Lepista nuda; Rhodopaxillus nudus.*

Etymology From Latin, "naked," from its light cuticle.

Description Cap 4–12 cm, convex, almost flat, rounded at edge when young, moist, smooth, sometimes with deep folds (spout-shaped), pale reddish brown or violet, more or less tawny-shaded in dry weather, fading with age. Gills thin, crowded, rounded toward stipe, fairly decurrent, light violet then brownish violet. Stipe 5–10 × 1–2 cm, sturdy, fibrous, solid, with enlarged fleecy base with lots of mycelial hyphae, lilac or lilac-gray, covered with light, floccose fibrils. Flesh violet-white, fragile in cap, fibrous in stipe. Odor pleasant, flavor sweetish. Spores transparent under the microscope, faintly lilac or pinkish in mass, elliptical, minutely roughened, 6–8 × 3–4 microns.

Edibility Excellent cooked (parboiling recommended), slightly poisonous raw.

Habitat In humus-rich ground in broadleaf and coniferous woods, sometimes on sawdust or in grassland.

Season Usually in autumn, but also at other times of the year except during periods of frost.

Note This is common throughout North America, easily recognized, and a popular edible.

Caution Slightly poisonous when raw.

75   LACCARIA AMETHYSTINA

Etymology   From Latin, "amethyst-colored."
Description   Cap 1.5–8 cm, initially convex and fairly umbilicate then expanded, depressed, sometimes perforated at the center, margin thin, somewhat irregular when mature, often split, slightly striate, dark violet or discolored depending on age and state of saturation of the cap, finely feltlike-squamulose. Gills distant, violet, adnexed-decurrent. Stipe 4–10 × 0.4–1 cm, cylindrical in young specimens, then compressed, fistulous, tomentose at halfway level, hairy at base. Flesh thin, hygrophanous, elastic, pale lilac-violet. Slightly fruity odor, sweet flavor. Spores white, globose, aculeate, 8.5–9 microns.
Edibility   Good.
Habitat   In coniferous and broadleaf woods, and in scrub.
Season   Summer and late autumn.
Note   This is an easily identifiable fungus because of the pale lilac-violet coloration of the carpophore. *L. bicolor* has a reddish cap but pale lilac gills and stipe.

---

76   LACCARIA LACCATA

Etymology   From Persian, "painted" or "varnished."
Description   Cap 1.5–5 cm, pale pinkish if wet or moist, ochreous when dry, convex then unevenly flat, fairly umbilicate-depressed, moist becoming dry, cuticle thick, broken up into small mealy scales, sometimes silky, undulate-wrinkled when mature. Gills flesh-colored, powdery when mature, adnexed-decurrent, broad, and distant. Stipe 7–10 x 0.6–1 cm, cap-colored, tough, fibrous, often supple, base whitish and hairy, fibrillose and striate, solid then hollow at cap attachment. Flesh cap-colored, thin, quite tough in stipe. Insignificant odor and flavor. Spores white, globose, aculeate, 8–9 microns.
Edibility   Fair.
Habitat   Common in fresh parts of woods.
Season   Summer and late autumn.
Note   There are many similar species, and a microscope is necessary for identification. We should mention *L. proxima* which prefers marshy or boggy areas; the cap is relatively large, orange, and squamulose; *L. tortilis*, a small fungus growing in damp or wet places with a striate, translucent cap; and *L. striatula* which lives in peat bogs.

77   LEUCOCORTINARIUS BULBIGER

Etymology   From Latin, "bulbed."
Description   Cap 5–10 cm, light brick-yellow, often with velar remains, especially at the edge, in the form of cobwebby or slightly membranous brownish scales, on a brown-ochreous background, with pinkish shading, or reddish brown or leather-colored, at first hemispherical, moist, turning almost flat. Gills white then cream-colored, dirty reddish brown, barely adnate, quite crowded. Stipe 5–10 × 1–1.2 cm, enlarged at the base in the form of an abrupt bulb, up to 2.5–3 cm wide, marginate, and with rather ring-shaped remains of a cobwebby veil, thick, whitish, occasionally with blackish fibrils, solid. Flesh white, reddish beneath cuticle and at cap attachment. No particular odor or flavor. Spores white, elliptical, roughened, 6–7 × 4–5 microns.

Edibility   Quite good.
Habitat   Coniferous woods in mountains.
Season   Late summer and autumn.
Note   The large conspicuous curtain in young specimens and certain other characteristics suggest a *Cortinarius*, although this species has white, not rusty-ochre, spores. This species has not been reliably reported in North America.

---

78   CLITOCYBE PHYLLOPHILA

Synonym   *Clitocybe cerussata*.
Etymology   From Greek, "leaf-loving."
Description   Cap 5–10 cm, convex then flat-depressed, covered with a silky white layer, especially near the edge, which conceals an ochreous, cream-colored background appearing with age in the form of uneven speckling. Gills white, turning yellowish or pinkish cream, adnexed-decurrent, fairly distant. Stipe 5–8 × 0.5–0.8 cm, cap-colored, elastic, fibrous, cylindrical, hairy at the base, often curved. Flesh white, thin. Strong odor of rancid meal, flavor similar. Spores whitish, oval or subglobose, smooth 4–5 × 3–4 microns.

Edibility   Poisonous.
Habitat   In groups on leaves, especially of beech.
Season   Midsummer to late autumn.
Note   This species may be confused with *Clitopilus prunulus* **(162)**, an edible pink-spored mushroom with long, decurrent gills. Ingestion causes muscarine poisoning.

79   CLITOCYBE FLACCIDA

Etymology   From Latin, "limp," from flesh texture.
Description   Cap 5–8 cm, rust-ochre, reddish, or reddish brown, shiny, not very fleshy, quite tough, limp when dry, always umbilicate then funnel-shaped, smooth, rarely squamulose. Gills whitish tending to turn yellow especially at the ends, decurrent, very curved, crowded, straight. Stipe 2.5–5 × 0.5–1 cm, reddish brown, rust-colored, elastic and tough, solid, cylindrical, enlarged and hairy at base. Flesh pale, thin, fragile, then limp when dry. Cyanic odor and sweetish flavor. Spores white, globose, minutely warty, 3–4 microns.
Edibility   Good.
Habitat   Gregarious, often in rings, in broadleaf woods.
Season   Summer and autumn.
Note   Two similar yellowish fungi with the same cap and gills grow beneath conifers: *C. vernicosa,* isolated, with a nonumbilicate cap but with curved margin, shiny; and *C. splendens,* which grows in groups and has forked gills. Similar but microscopically different fungi include: *C. sinopica,* conspicuous red or reddish brown, with a preference for coniferous burn sites; and *C. vermicularis,* flesh-red if wet, with white mycelial threads at the base.

---

80   CLITOCYBE GEOTROPA

Etymology   From Greek, "erect," because of its stance.
Description   Cap 8–30 cm, chamois-brown tending to turn pale, fleshy, sometimes rather limp, broadly funnel-shaped with a central umbo, dry, downy at the center, often squamulose, margin involute, pubescent. Gills whitish, long and decurrent, pointed at the tips, quite crowded, soft. Stipe 5–10 × 1–4 cm, whitish, fibrillose-striate, elastic, narrowing toward the top, solid then pithy. Flesh whitish, thick, soft. Smell aromatic, flavor sweetish. Spores white, elliptical, smooth, 4–6 × 3–4 microns.
Edibility   Very good when young.
Habitat   In groups, sometimes very large ones, in woodland and grassland.
Season   Summer to late autumn.
Note   *C. maxima* has a stipe smaller than the diameter of the cap, which is not umbonate at the center and has a flocculose cuticle.

81 CLITOCYBE CLAVIPES

Etymology From Latin, "club-footed."
Description Cap 4–6 cm or more, two-layered, ash-gray, sometimes whitish at the margin, rarely white all over, fleshy, slightly convex then flat, becoming eventually slightly concave, sometimes umbonate, smooth, although under the microscope it shows a fine covering of fibrils. Gills white, sometimes yellowish, long and decurrent, soft. Stipe 3.5–10 × 0.6–1.2 cm, at the base up to 4 cm, markedly club-shaped, soft, spongy, tough, sometimes fibrillose. Flesh ash-gray then white, soft, thin at margin of cap. Pleasant smell, quite specific, flavor sweetish. Spores white, elliptical, smooth, 5–7 × 3–4 microns.
Edibility Mediocre, but see **caution.**
Habitat In broadleaf and coniferous woodland.
Season Summer and autumn.
Note This species is quite distinctive because of its club-shaped stipe which is soft and spongy. Some forms of *C. nebularis* (83) and *C. alexandri* may be similar.
Caution Should not be eaten with or followed by alcoholic beverages: some people develop a headache and transient upper body rash, a "poisoning" similar to but not the same as that caused by the inky cap *Coprinus atramentarius.*

82 CLITOCYBE LANGEI

Etymology After the Danish mycologist Lange.
Description Cap 1–3 cm, gray-brown tending to turn hazel when drying, or chamois- or cream-colored, depressed, glabrous when mature, in young specimens covered with a fine pruinescence, almost a whitish feltlike down which persists towards the edge, striate when moist. Gills adnexed, slightly decurrent, gray or yellowish, unequal. Stipe cylindrical, grayish brown, then light brown, covered first with fine sandy hairs which give it a whitish and hoary appearance, then slightly fibrillose, elastic, solid. Flesh fine, hygrophanous. Strong fragrant odor and flavor of meal. Spores white, piriform, smooth, 5–6.5 × 2.7–3.2 microns.
Edibility Good.
Habitat A northern species, often in groups, beneath conifers and birch.
Season Late summer to late autumn.
Note This is a small aromatic fungus, not unlike *C. pausiaca* and *C. vibecina,* two fairly similar species with the same habitat. None of these species is known to occur in North America.

83 CLITOCYBE NEBULARIS

Etymology From Latin, "mist-gray."
Description Cap 7.5–20 cm, light, brownish, or dark gray, rarely white, sometimes as if covered with a whitish pruinescence but in fact glabrous, smooth, thick, convex then flat, eventually slightly depressed at disc. Gills whitish, sometimes with yellowish shades, short and decurrent, curved, crowded, thin. Stipe 7–12 × 2–3 cm, whitish, enlarged at base, narrowing at top, spongy and elastic. Flesh white and thick. Strong odor and flavor. Spores yellowish, elliptical, smooth, 7–8 × 3–4 microns.
Edibility Good when cooked. **May cause allergic upset.**
Habitat Common in woodland, grassland, and moorland, in groups.
Season Summer and autumn.
Note In Europe, this species is quite sought after because of the strong aroma of the flesh; in North America, however, the odor is typically unpleasant, and the mushroom, which is mostly western in distribution, is generally not eaten. *C. alexandri* is a similar fungus which grows in woods; it has a brownish cap with gray or ochre shading, finely pubescent, with a club-shaped base with a mycelial tuft that envelops the substratum and leaves, and ochreous, light brown gills.

84 CLITOCYBE ODORA

Etymology From Latin, "perfumed."
Description Cap 3–9 cm, typically greenish to off-white; it fades on drying and is paler growing under leaves, fairly finely fibrillose, from convex to flat, rarely depressed, or slightly umbonate, margin incurved, pubescent. Gills white to green-tinged, longer than the flesh of the cap, quite distant, adnexed-decurrent. Stipe 3–5 × 0.6–0.8 cm, cap-colored, cylindrical, supple, floccose-fibrillose then smooth, hairy and enlarged at the base. Flesh dirty white, elastic. Very strong odor of aniseed, similar flavor. Spores white, elliptical, smooth, 6–8 × 3–4.5 microns.
Edibility Good.
Habitat In woods, usually broadleaf.
Season Summer and autumn.
Note Its presence is often given away by the strong odor of aniseed. Other species have the same odor, although to different degrees: *C. subalutacea,* pale yellowish; *C. obseleta,* tuft-like; *C. fragrans,* whitish, nonumbilicate cap, growing in moss; *C. suaveolens,* very similar to the previous species with an umbilicate cap, darker at the center, with margin striate.

85 ARMILLARIELLA MELLEA

Synonym *Armillaria mellea.*
Common name Honey fungus.
Etymology From Latin, "like honey," from its color.
Description Cap 3–15 cm, globose becoming depressed, slightly rounded, cuticle brownish ochre with green and red shades, disc covered with erect brown hairs spreading out toward the edge. Gills quite crowded, adnate, decurrent or non-decurrent, whitish then yellowish, eventually with reddish spots. Stipe 10–20 × 1.5–5 cm, long club-shaped, often curved, color variable, darker toward the base, similar to cap color, white ring, striate at top, yellowish and floccose at bottom. Flesh fine, fibrous in stipe. Oily odor, quite unpleasant, flavor somewhat bitter, slightly astringent. Spores white, elliptical, smooth, 7–9 × 5–6 microns.
Edibility Good when cooked, suitable for drying. **Somewhat poisonous: parboiling is recommended.**
Habitat Parasitic or saprophytic on wood, but also symbiotic, cosmopolitan, typically clustered.
Season Summer and autumn.
Note Because of its variability and the number of poisonous look-alikes, confirm all identifying characters. Differing only in its lack of a ring, *A. tabescens* is equally edible, given the above precautions.
Caution Somewhat poisonous. See note.

86 OMPHALOTUS OLEARIUS

Synonyms *Clitocybe olearia; Clitocybe illudens.*
Etymology From Latin, "of the olive (tree)," because of its habitat on this host in Europe.
Description Cap 6–14 cm, convex then expanded and depressed, with radial innate fibrils, dry, from light orange-yellow to mahogany-brown, margin thin, often lobate and split. Gills very decurrent, crowded, unequal, orange-yellow, arcuate. Stipe 8–16 × 1–2 cm, central or eccentric, narrowing at base, striate-fibrillose, tough, cap colored. Flesh orange-yellow, staining, tough. Odor strong and oily, flavor astringent. Spores pale yellow-white, globose, smooth, 5–7 × 4.5–6.6 microns.
Edibility Poisonous.
Habitat At the foot of stumps or unhealthy trees, or on the ground on roots, parasitic on broadleaf species, often growing in tufts.
Season Late summer up to first winter frosts.
Note In the dark the gills of fresh fruit-bodies give off a greenish glow. Can be confused with the chanterelle (*Cantharellus cibarius*, **154**) which grows on the ground and has thick-edged forked ridges instead of gills, or with *Hygrophoropsis aurantiaca* which has clearly forking gills, both of which are edible.

87 LEUCOPAXILLUS GIGANTEUS

Synonym *Clitocybe gigantea.*
Etymology From Greek, "gigantic," because of its size.
Description Cap 30–40 cm or more, convex then flat, depressed, eventually funnel-shaped, soft to touch, white with minute adpressed scales, silky, turning slightly brown, margin markedly involute, first pubescent then revolute and smooth and grooved with sometimes pale ochre small depressions, rarely split at grooves. Gills whitish then slightly ochreous, subdecurrent, often ramified and anastomosing, crowded, detachable from flesh. Stipe 3–7.5 × 2–5 cm, cylindrical or narrowing at top, with enlarged base, smooth or slightly pubescent, white. Flesh white, solid. Pleasant, mealy odor, sweet flavor. Spores whitish, broadly elliptical, smooth, 7–8 × 5–6 microns, amyloid.
Edibility Very good.
Habitat Gregarious, often in rings, in fields and meadows, at the edge of woodland.
Season Summer and autumn.
Note *L. candidus* is smaller and its cap is never squamulose. *L. lepistoides* is large and grows in warm temperate areas; the cap becomes greenish at the center, then shaded with ochre, margin not grooved, strong, slightly aniseedlike odor, gills adnexed-decurrent.

88 PSEUDOCLITOCYBE CYATHIFORMIS

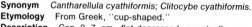

Synonym *Cantharellula cyathiformis; Clitocybe cyathiformis.*
Etymology From Greek, "cup-shaped."
Description Cap 2–7 cm, flat-depressed, eventually cup-shaped, sometimes undulate, ash-gray/brown or dark gray, fading when dry, shiny when wet, opaque when dry, very hygrophanous, fairly striate at margin. Gills ash-gray, grayish sometimes with reddish spots, adnate or decurrent, connate at stipe, rarely forked. Stipe 5–10 × 0.6–0.9 cm, grayish, elastic, narrowing toward top, fibrillose-reticulate, with white hairs at the base. Flesh watery, thin, grayish. Cyanic odor, sweetish flavor. Spores white, elliptical, minutely warty, 7.2–12 × 4–7.5 microns, amyloid.
Edibility Fair.
Habitat Especially on rotten wood; also in grass or moss in meadows, near trees and under bushes, often gregarious.
Season Late autumn, winter, and spring.
Note This genus includes two other species which are likewise gray with amyloid spores: *P. expallens,* with deeply umbilicate cap, markedly striate, gills pointed at tips; and *P. obbata,* with very dark cap and gills and much lighter brown stipe.

89 RICKENELLA FIBULA

Synonym *Omphalina fibula; Mycena fibula.*
Etymology From Latin, "pin-shaped."
Description Cap 0.4–2 cm, usually orange-yellow, tending to turn paler when drying, membranous, campanulate soon becoming umbilicate and eventually funnel-shaped, glabrous, striate when wet. Gills whitish or yellowish, very decurrent, distant. Stipe 3–8 × 0.1–0.2 cm, cap-colored, threadlike, wavy in appearance, often pubescent under the microscope. Flesh very fine, cap-colored. Spores white, elliptical, smooth, 3–4 × 2 microns.
Edibility Of no value because of size.
Habitat Typically in moss, sometimes in woods or meadows.
Season All year.
Note A very similar species, *Xeromphalina campanella,* is smaller, often gregarious, and grows on the bark of tree trunks and branches. *Rickenella setipes (Mycena swartzii)* has an ochre to reddish brown cap with a darker center, and a yellowish stipe, brown or dark violet at the top. *Omphalina cyanophylla,* initially ash-gray then yellowish, grows on rotting coniferous wood.

90 MYCENA HAEMATOPUS

Etymology From Greek, "with a bleeding foot."
Description Cap 2–4 cm, grayish with brownish and purple shades, tending to fade with age, dark at disc, conical then campanulate, obtuse, initially powdery then glabrous, with salient edge in young specimens, slightly striate. Gills white then pinkish or violet, adnate, with alternate lamellae. Stipe 5–10 × 0.2–0.5 cm, white, grayish, pinkish, or violet, eventually ash-gray, rigid but fragile, often curved at base, hairy, initially with a thick powdery pruinescence, short-lived, then smooth, producing droplets of blood-red latex on bruising. Flesh fine, turns red. No odor or flavor. Spores white, elliptical, smooth, 8–11 × 5–7 microns, amyloid.
Edibility Can be eaten, but quality bad.
Habitat In fairly large groups, tuftlike, on rotting wood, particularly birch or oak.
Season Summer, autumn, and spring.
Note The yield of colored or white latex is quite common in this genus. *M. sanguinelenta,* often terrestrial, is almost completely glabrous with a conspicuously salient edge. It is brownish, with various shading, and yields a red latex. The latex produced by *M. erubescens,* with its bitter flesh, ranges from colorless to whitish, sometimes becoming pinkish or wine-red.

91 MYCENA EPIPTERYGIA

Etymology From Greek, "surmounted by a small wing," from the pellicle of the cap.

Description Cap 1–3 cm, gray, ash-gray, or yellow, tending to whiten, campanulate, membranous, striate, covered with a viscid transparent pellicle, completely separable, margin often denticulate. Gills white and adnate. Stipe 5–8 × 0.1–0.2 cm, cap-colored, also covered with a viscid, separable pellicle, small rhizomorphs at base. Flesh fine, whitish. No particular smell or flavor. Spores white, elliptical, smooth, 8–12 × 5.5–7.5 microns, amyloid.

Edibility Of no value because of size.

Habitat In woodland and grassland, on leaves and detritus from trees, in moss.

Season Summer and autumn.

Note *M. viscosa* grows beneath conifers, among the needles or on stumps, has reddish markings on the cap, and emits a strong smell of rancid fat. *M. epipterygioides* has greenish coloration on the stipe, the pale yellowish cap, and the gills.

---

92 MYCENA INCLINATA

Etymology From Latin, "not straight" or "bent."

Description Cap 2–3 cm, gray-brown, quite dark, globose, campanulate, obtuse, eventually flat or slightly depressed, smooth, shiny in dry weather, margin striate, fairly salient, slightly crenate. Gills whitish, grayish at attachment, sometimes pinkish, adnate, crowded. Stipe 6–10 × 0.2–0.4 cm, whitish or brownish becoming orange-yellow from base upward, eventually reddish brown at base of old specimens, initially elastic and leathery then fragile, slightly pruinous and fibrillose, with hairy base. Flesh whitish in the cap, light ochre-brown in the stipe. Slightly alkaline odor, flavor slightly astringent, rancid. Spores white, elliptical, smooth, 7–10 × 5–7 microns, amyloid.

Edibility Caps edible, but quality inferior.

Habitat In dense tufts on stumps and branches of broadleaf species, especially oak and beech.

Season Spring and autumn.

Note Many similar species grow clustered on wood but they lack a crenate cap margin or have a stipe that is smooth and contains latex.

93 MYCENA GALERICULATA

Etymology From Latin, referring to a particular sort of headgear worn in ancient times.

Description Cap 2–5 cm, brownish gray but varying to yellow or blackish red, sometimes white, conical-campanulate then expanded and striate to the umbo, dry and smooth. Gills whitish then pinkish, adnate-decurrent, sometimes connected by veins. Stipe 5–12 × 0.3–0.5 cm, cap-colored, smooth, fragile when mature, often curved with hairy base, white, and rooting. Flesh fine, grayish. Mealy odor and flavor. Spores white, broadly elliptical, smooth, 10–11 × 6–8 microns, amyloid.

Edibility Can be eaten, but of little interest.

Habitat Tufted, with the base of stipes connected by hairs, on coniferous and broadleaf stumps and trunks.

Season All year.

Note *M. polygramma*, growing on wood, has a bluish gray conspicuously grooved and striate stipe. *M. alcalina*, also forming tufts on dead wood, resembles *M. galericulata*: it is brownish-gray with ash-gray (not pinkish) gills and a strong alkaline odor. This odor and a carpophore with reddish-brown markings especially on the gills, are also typical of *M. maculata*, which is more gregarious than tufted, with base of stipe often rooting if the wood is soft and decomposed.

94 MYCENA PURA

Etymology From Latin, "clean."

Description Cap 2–8 cm, purple to pale lilac, campanulate then expanded, eventually flat, sometimes slightly umbonate, margin striate. Gills cap-colored but paler, adnate, broad, ventricose, connected by veins. Stipe 3–10 × 0.2–1 cm, cap-colored, tough, leathery, hollow, narrowing at top, smooth with fibrils running lengthwise, white and hairy at base. Flesh whitish. Sometimes odor and flavor of radish. Spores white, elliptical, smooth, 6–9 × 2.5–4 microns, amyloid.

Edibility Edible in small quantities.

Habitat Gregarious in woodland, in leaves and moss.

Season Summer and autumn.

Note There are many varieties of *M. pura*, all identifiable by the strong smell of radish and the fact that the gill edges are not blackish. The variety *alba* is entirely white; *lutea* has an ochreous-yellow cap and violet stipe; *multicolor* has a greenish blue-gray cap with a yellow umbo and pinkish purple stipe; *rosea* is sturdier with a pink cap and paler or whitish stipe. A similarly radish-scented species, *M. pelianthina*, is entirely violet or purple, and has dark purple-brown gill edges.

95 MYCENA VULGARIS

Etymology From Latin, "common."

Description Cap 0.6–1 cm, fairly dark grayish brown, sometimes whitish except at the center, or reddish ochre in old specimens, campanulate then convex, depressed at the center with papilla, slightly striate, covered with a thin viscid pellicle, separable. Gills white or gray, slightly decurrent, thin. Stipe 2.5–5 × 0.1–0.2 cm, ash-colored, very viscid, with rooting base with white hairs. Flesh whitish. No special odor or flavor. Spores white, elliptical, smooth, 6–9 × 3–4 microns, amyloid.

Edibility Of no value because of size.

Habitat Gregarious, grows in large numbers in needle litter in coniferous woods.

Season Summer to late autumn, including period of first winter frosts.

Note Other species of *Mycena* occur in coniferous woods in winter and early spring: *M. strobilicola* grows on cones of Norway spruce; *M. seynii,* found on maritime pine cones, has a very hairy base.

96 MYCENA ROSELLA

Etymology From Latin, "small rose."

Description Cap 0.7–1 cm, membranous, campanulate then hemispherical, obtusely umbonate, slightly hygrophanous, pinkish, striate. Gills pinkish, adnate, quite distant, edge darker blackish purple. Stipe 2–3.5 × 0.1 cm, soft, pinkish with white, tomentose base. Flesh very fine, white, reddish in stipe. No particular smell or flavor. Spores white, elliptical, smooth, 6.5–10 × 4–5 microns, amyloid.

Edibility Of no interest because of size.

Habitat Gregarious, on conifer needles.

Season Late spring to late autumn.

Note A principal characteristic of the species is the fact that the color of the gill edge differs from the rest of the fungus. Other small fungi with this feature include: *M. citrinomarginata,* found on wood, with yellow gill edges; *M. flavescens,* with a smell of radish; *M. chlorantha,* with an iodized smell; *M. elegans* has bright orange-red gill edges. Many species have darker gill edges, including: *M. avenacea,* with a brown cap; *M. albidolilacea,* with a whitish cap with pale lilac-pink shading; *M. atromarginata,* which is entirely brownish-gray; *M. capillaripes,* found in needles, with an alkaline odor; and *M. viridimarginata,* with greenish gill edges.

97 HYGROPHORUS LIMACINUS

Etymology From Latin, ''mucous.''
Description Cap 5–10 cm or more, olive-gray to soot-brown, darker at disc, convex then expanded, slightly raised at center, covered with a separable glutinous pellicle, cuticle also detachable, shiny in dry weather, glabrous or with innate fibrils. Gills white with grayish or pale yellowish shading, distant, adnate-decurrent. Stipe 5–8 × 1–1.5 cm, sturdy, ventricose, sometimes with pointed base, viscous and at the same time floccose-squamose, grayish in lower half, white and squamulose at top. Flesh white, soft. No particular smell or flavor. Spores white, elliptical, smooth, 9–13 × 5.5–7.5 microns.
Edibility Very good if cuticle is removed.
Habitat In woodland in leaves and moss, prefers pine.
Season Summer and autumn.
Note *H. olivaceoalbus* is more slender, and grows in spruce and fir woods. In North America we also find *H. paludosus*, mostly in the Southeast. It is similar to both the above when young, but with age, it takes on purplish or pinkish coloration in the cap, and becomes marked with green at top of stipe and on gills.

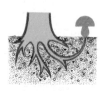

98 HYGROPHORUS MELIZEUS

Etymology From Greek, ''honey-colored.''
Description Cap 2.5–4 cm, very light leather-yellow, pale hazel-ochre, uniformly colored, at least in central area, very viscid, convex then flat with raised edge, margin at first pubescent. Gills yellowish cream-colored, decurrent, distant, connected by veins. Stipe 7–8 × 0.6–1 cm, cap-colored, narrowing toward base, conspicuously flocculose and white at the top in wet weather, dry, nonglutinous even if wet. Flesh light cream-colored. Inconspicuous odor and flavor. Spores whitish, ellipsoid, smooth, 6.5–9 × 4.5–5 microns.
Edibility Can be eaten.
Habitat Gregarious, in grass usually beneath birch.
Season Late summer and autumn.
Note Not known in North America, this fungus is often confused with *H. flavodiscus* (the cap is almost identical but the latter's stipe is viscid, cream-colored, the gills are whitish) and with *H. cossus* (which has a white cap turning cream-colored at disc, with often pale ochre gills and a glutinous stipe with watery droplets that turn yellowish as they dry out).

99 HYGRAPHORUS AGATHOSMUS

Etymology   From Greek, "nice-smelling."
Description   Cap 4–7 cm, uniform pale brownish gray, often viscid, especially at center covered with small raised papillae, crowded, transparent white, convex then flat, humped, margin first involute, hairy, eventually spreading and undulate. Gills pure white, decurrent, distant, soft, sometimes venose at base. Stipe 5–12 × 0.6–1.5 cm, white, cylindrical or slightly enlarged at base, sometimes striate with fibrils, with mealy granulations at top tending to turn brownish-gray, solid, full. Flesh whitish, soft. Strong odor of celery with a hint of aniseed or bitter almonds, flavor sweetish. Spores white, broadly oval, smooth, 8–10 × 4–6 microns.
Edibility   Good, although the smell is not to everyone's liking.
Habitat   Grows among spruce and pine.
Season   Summer and autumn.
Note   *H. pustulatus,* brownish-gray with darker cap disc, is umbonate and normally covered with papillae; top of stipe has black spots; the gills, sometimes with pale green highlights, are adnexed-decurrent; no particular odor. *H. tephroleucus,* with tomentose cap edge, resembles *H. pustulatus,* but has stipe entirely covered with black speckling.

100 HYGROPHORUS CAPRINUS

Etymology   From Latin, "pertaining to goats."
Description   Cap 3–10 cm, sooty black or blackish, with pale bluish shades, convex then flat and depressed, at times fairly umbonate, especially when immature, solid, moist or dry, with fairly radial, innate, blackish fibrils, pellicle separable only at margin, at first white, pruinous and involute, then expanded and at times recurved, same color as the rest of the cap, and undulate. Gills white, tending to turn pale green or gray, very decurrent, distant, often connected by veins. Stipe 4–8 × 1–1.5 cm, sooty gray, cylindrical or narrowing at the base, whitish at top, full. Flesh white. Strong indefinable odor, sweet flavor. Spores white, elliptical, slightly pointed at one end, smooth, 6–9 × 4–5 microns.
Edibility   Very good.
Habitat   In heathland and meadows beneath fir.
Season   Summer and autumn.
Note   *H. camarophyllus* is very similar but does not have any bluish coloration on the cuticle. The gills are whitish turning grayish cream-colored. Other species are *H. calophyllus,* gray, with a dry stipe, cap viscous and nonfibrillose, gills typically pinkish; and delicious *H. marzuolous* (102).

101 HYGROPHORUS RUSSULA

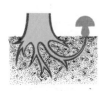

Etymology From Latin, "reddish."
Description Cap 10–20 cm, flesh-colored or pale purple with more deeply colored speckling, paler and whitish at tomentose margin, slightly viscid, humped, flat-convex, eventually depressed, slightly undulate. Gills whitish, often with deep red markings, sinuate to adnexed becoming decurrent, thin, quite crowded. Stipe 6–12 × 1–2 cm, white, spotted red, with the upper part white, mealy, solid, fairly cylindrical, often curved. Flesh white, compact, fibrous in stipe. No particular odor or flavor, sometimes slightly bitter. Spores white, elliptical, smooth, 6–8 × 4–5 microns.
Edibility Good.
Habitat Beneath broadleaf trees, especially oak.
Season Late summer and autumn.
Note Very common in warm temperate regions, less so farther north. A similar species is *H. erubescens,* which prefers conifers. The flesh is pale yellow, slightly bitter; the decurrent gills are, at most, suffused with pale flesh-pink; the stipe is red but takes on a yellow coloration when touched; the viscid cap has pink markings.

102 HYGROPHORUS MARZUOLUS

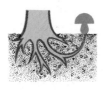

Etymology From Latin, "of March."
Description Cap 3–10 cm, fairly dark ash-gray, sometimes with ochreous shading, becoming speckled blackish gray with age, at first convex, soon becoming flat and depressed, irregularly humped, with margin expanded, wavy-lobate, cuticle initially moist, drying out rapidly. In young specimens gills are crowded, large, short, distant; they then become thinner, arcuate, decurrent, connected by veins and ramified, white tending to turn gray or blackish. Stipe 4–8 × 1.2–3 cm, squat, full, cylindrical, straight or curved, or tapered at base, furfuraceous, white at top, otherwise silvery gray, solid. Flesh thick, white, faintly gray beneath cuticle, tender, slightly fibrous in base. No particular odor or flavor. Spores white, elliptical, smooth, 7–9.5 × 5.5–6.5 microns.
Edibility Excellent.
Habitat Gregarious, beneath litter of leaves and moss in coniferous and broadleaf woods, especially in mountainous areas.
Season Late winter, early spring, rarely in late autumn.
Note Its distribution is generally limited to central Europe and North Africa; in North America it has been reported near melting snowbanks in the Pacific Northwest.

103 CAMAROPHYLLUS VIRGINEUS

Synonym *Hygrophorus virgineus.*
Etymology From Latin, "virginal," because it is snow-white.
Description Cap 3–7 cm, convex, obtuse, flat with a slight umbo, eventually depressed, cuticle moist, broken up into small adpressed scales, floccose, turning yellow when drying out. Gills white, decurrent, distant, quite thick, joined by veins at base. Stipe 5–11 × 0.5–1 cm, white, solid, full, pruinous and striate, narrowing at base, sometimes reddish. Flesh white, slightly fibrous. No particular odor or flavor. Spores white, elliptical, smooth, 8–12 × 5–6 microns.
Edibility Good.
Habitat In grassland and woods, often gregarious.
Note Another entirely white species, *Hygrophorus niveus,* is very similar, with a thin, fairly striate cap, yellowish or brown shading in old specimens, and a grassy odor. *H. russocoriaceus,* which is at first dirty white and then lighter-colored as it dries, has a distinctive odor of leather.

104 HYGROPHORUS EBURNEUS

Etymology From Latin, "ivory-white."
Description Cap 3–10 cm, pure white, shiny, disc at most cream-colored or faintly yellowish, or else entirely cream-colored, or the color of pale leather, never darker, very viscous, convex-flat with the peripheral part turned outward when mature, margin at first involute, pubescent. Gills white, decurrent, distant, venose at the base. Stipe 3–8 × 0.6–1.5 cm, white, not always regular, smooth, viscid except at the top, floccose, emitting small lactescent droplets which do not turn yellow as they dry out. Flesh white, constant. Strong distinctive odor (which lingers on the fingers) of sage, flavor similar. Spores white, elliptical, smooth, 8–10 × 4–5 microns.
Edibility Can be eaten, but of poor quality.
Habitat Gregarious, in broadleaf woods and grassland.
Season Summer and autumn.
Note A very similar species, *H. chrysodon,* grows in coniferous woods in mountainous areas; it forms more slender carpophores, with a pure white cap not very viscid, becoming cream-colored as it dries out. The flesh has no specific odor, merely mushroomy. Its flesh recalls that of *H. eburneus,* but the carpophores are easily identifiable because the top of the stipe and the margin of the cap are yellowish and floccose.

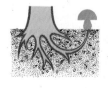

105 HYGROCYBE SPADICEA

Synonym *Hygrophorus spadiceus.*
Etymology From Greek, "date-brown."
Description Cap 1–6 cm, olive-ochreous-brown, blackish and shiny when dry, fragile, campanulate, but expanded, acute or obtuse, with radial innate fibrils, viscid in wet weather. Gills lemon-yellow, broad, distant, quite sinuate, ventricose, adnate or almost free. Stipe 4–7 × 0.6–1 cm, yellowish, dry, striate with brown or lemon-yellow fibrils, cylindrical. Flesh pale lemon-yellow. No special odor or flavor. Spores white, elliptical, smooth, 8.5–12.5 × 4.5–7.5 microns.
Edibility Mediocre.
Habitat In moist grassland with moss, or near broadleaf trees.
Season Summer and autumn.
Note Although rare in North America, this is one of the most easily indentifiable of the *Hygrocybe* group because of the sharp contrast in color between the cap and the yellow gills and stipe. *H. calyptraeformis* is also easily recognizable. It grows in meadows and woodland; the cap is conical, acute, pinkish with innate fibrils, or pale lilac-mauve, as are the gills; the stipe is up to 12 cm long, white, or with pale lilac shading.

---

106 HYGROCYBE MINIATA

Synonym *Hygrophorus miniatus.*
Etymology From Latin, "red lead-colored," after a lead-based antirust paint.
Description Cap 0.5–2 cm, crimson-red, paling with age and becoming opaque, convex, often umbonate and eventually umbilicate, glabrous or squamulose. Gills yellow or orange-yellow, adnate, distant. Stipe 3–5 × 0.2–0.4 cm, crimson-red, shiny, fibrillose, cylindrical. Flesh waxy, red. No special odor or flavor. Spores white, elliptical, smooth, 6–7 × 4–5 microns.
Edibility Can be eaten but of little interest.
Habitat Typically in moss.
Season Late spring, summer, and autumn.
Note Although it is small, this fungus is easily spotted because of the bright color which stands out sharply from the surrounding environment. *H. cantharellus* has a squamulose cap and decurrent gills, color ranging from whitish to pale yellow. *H. turundus* has a markedly flocculose cap and grows in sphagnum moss in peat bogs. *H. coccineus* and *H. reai,* two more red species, with glabrous caps, tend to fade to orange and yellow with age. The latter species usually has bitter flesh.

107 HYGROCYBE PUNICEA

Synonym *Hygrophorus puniceus.*
Etymology From Latin, "of pomegranate" or "garnet-red."
Description Cap 5–11 cm, scarlet-red tending to turn pale
with age starting from the center, campanulate, obtuse, margin
normally involute, lobate, viscid. Gills yellow, often red at base,
ascending and apparently free, ventricose, broad, thick, and
distant. Stipe 7–11 × 1–2.5 cm, cap-colored or bright yellow,
base invariably white, fusiform, often curved, fibrillose, striate,
squamulose at top, solid, quickly becoming hollow. Flesh cap-
colored, initially white, slightly watery, waxy, slightly fibrous in
the stipe. No particular odor or flavor. Spores white, broadly
elliptical, smooth, 8.5–11 × 5–5.5 microns.
Edibility Good.
Habitat In woods under broadleaf trees and conifers.
Season Summer and autumn.
Note This fungus is one of the largest and brightest members
of its genus. *H. coccinea,* also red but turning orange with age,
has a stipe that is almost smooth and nonstriate. Most of the
many red, orange, or yellow species are small or smallish in
size.

108 HYGROCYBE PSITTACINA

Synonym *Hygrophorus psittacinus.*
Etymology From Greek, "parrot," because of the colors.
Description Cap 2–5 cm, initially olive-green, bright light
bluish green, shiny, then yellowish, pale yellow, finally brick-
colored before turning purple-brown, campanulate then partially expanded, umbonate, striate, viscid. Gills yellow, greenish at base, adnexed, ventricose, quite distant. Stipe 4–7 ×
0.4–0.7 cm, cylindrical, bright green because of the mucus,
which eventually only persists at the top, otherwise turning yellowish. Flesh thin, waxy, watery, white shaded with green or
yellow. No particular odor or flavor. Spores white, elliptical,
smooth, 8–9 × 4–5.7 microns.
Edibility Mediocre.
Habitat In woods and grassland.
Season Summer and autumn.
Note *H. sciophana* has different red shading, with pale
greenish markings and yellow-edged gills. *H. laeta* also forms
carpophores with different colors. The cap and stipe may be
yellow, orange, salmon-pink, or pinkish, although the very top
may be pale lilac or greenish. The decurrent gills are salmon-
colored then pale lilac.

109 HYGROCYBE BREVISPORA

Synonym *Hygrophorus brevisporus.*
Etymology From Latin, "with short spores."
Description Cap 3–6 cm, yellow, pale greenish yellow, conical, campanulate, more open at the unevenly lobate margin, central area with pointed mamelon, sulcate-striate (radially), dry, at most, damp. Gills whitish then pale yellow, broad, ventricose, quite distant, almost free. Stipe 4–6 × 0.8–1.2 cm, cap-colored, soon becoming hollow, compressed, with lengthwise grooves, often forming fissures, briefly floccose at top, with orange shading at base, not viscous. Flesh pale yellow, watery, fibrillose, but fragile in stipe. No particular odor or flavor. Spores white, oval or globose, smooth, 6–8.5 × 5–6 microns.
Edibility Mediocre.
Habitat In grassland and grassy areas in woodland.
Season Late summer and autumn.
Note This fungus is not known in North America. Other species are very similar: *H. citrinovirens*, which has a slightly more slender stipe, 6–10 × 0.3–1 cm; *H. obrussea*, with a conspicuously compressed stipe, sulcate, up to 3 cm maximum width, smooth all over, dry, cap golden yellow, and yellow gills with white edge; and *H. chlorophana*, all yellow, with very viscid stipe.

110 HYGROCYBE CONICA

Synonym *Hygrophorus conicus.*
Etymology From Latin, "cone-shaped."
Description Cap 3–6 cm, scarlet, yellow, hazel, yellow-green, gray-brown, or sooty gray and yellow, tending to blacken all over, campanulate, quickly becoming conical with a pointed apex, often lobate and split when expanded, viscid when wet, silky when dry. Gills white or yellow, sometimes reddish at base, becoming black, semifree, ventricose, thin, crowded. Stipe 6–9 × 0.4–0.9 cm, red, yellow, or pale green, turning black, cylindrical, striate-fibrous, rigid and straight. Flesh thin, watery, same color as outside surface, blackening. No particular odor or flavor. Spores white, broadly elliptical, smooth, 9–14 × 4–8 microns.
Edibility Rather mediocre.
Habitat In grassland, woods, and by the roadside.
Season Summer and autumn.
Note This initially bright fungus changes color with age, and immediately after being picked. Similar species which don't turn black are: *H. mucronella*, cap red, stipe yellow turning red; *H. crocea*, lemon-yellow or orange all over, with fibrillose striate stipe; *H. intermedia*, cap downy and red, stipe yellow and fibrillose.

111 PLEUROTUS ERYNGII

Etymology From the name of the host plant.

Description Cap 5–10 cm, reddish brown to dark brown or ochreous-brown, slightly squamulose, striped with radial fibrils, adpressed, grayish brown, eventually glabrous, convex then flattened, faintly depressed, sometimes irregular, eccentric or concentric, margin thin, recurved. Gills fairly crowded, rarely ramified, very decurrent, whitish then ochreous-gray, edge darkening with age. Stipe 4–6 × 1–2 cm, usually eccentric, but sometimes central, cylindrical or narrowing and rooting at the base, sometimes curved, full, soft, elastic, fibrillose lengthwise, whitish, eventually ochreous-gray, with brownish mycelium at the base. Flesh thick, not hygrophanous, solid, white. Slight, indistinct but pleasant odor and sweet flavor. Spores white, elliptical, smooth, 10–12.5 × 5–5.5 microns.

Edibility Excellent.

Habitat On leaves of plants in the genus *Eryngium* and some related genera in the parsley family.

Season In warmer regions from spring onwards, then late summer and autumn.

Note *P. ferulae,* just as delicious, often grows in tufts at the bottom of giant fennel. Tasty *P. opuntiae* grows on prickly-pear cactus, agave, and yucca.

---

112 PLEUROTUS DRYINUS

Etymology From Greek, "of oaks."

Description Cap 5–10 cm, lateral, oblique, convex then quite flat, whitish, covered with dirt-brown down, often lighter, which breaks up into spotlike adpressed scales, becoming grayish brown, sometimes with edge appendiculate because of membranous and downy velar remains. Gills white, yellowish with age, decurrent with slight edge on stipe, often anastomosed, but never joining together at base, short and single. Stipe 2.5–10 cm, solid, full, tough, eccentric, white, almost wooly, downy, squamulose, with tapered base, at times with ringlike velar remains. Flesh white, thick, solid. Pleasant odor and flavor, like cultivated field mushrooms. Spores white, extended, cylindrical, smooth, 12–13 × 3–4 microns.

Edibility Good.

Habitat On trunks, living or dead, of broadleaf trees, particularly oak, maple, willow, walnut, and alder.

Season Late summer, autumn, and during the winter.

Note *P. calyptratus,* a rare species growing on poplars, with a fairly small lateral stipe, has a membranous, mucous veil which forms a braceletlike ring beneath the gills.

113　PLEUROTUS OSTREATUS

Common name　Oyster fungus, oyster mushroom.
Etymology　From Latin, "oyster," because of cap shape.
Description　Cap 6–14 cm, often imbricate, superposed, violet-black to brownish gray, fading with age, eccentric and asymmetrical, shell- or spatula-shaped, slightly depressed at attachment to stipe, smooth, shiny and glabrous. Gills initially creamy white, then ivory-white, long, fairly crowded, unequal, decurrent along stipe. Stipe 2–8 × 1–2 cm, white, smooth, full, oblique, lateral, rarely central, enlarged at top, rudimentary, hairy base. Flesh white, thick, at first tender then slightly tough. Fairly pleasant odor, tasty flavor. Spores pale cream-colored or shaded with pale lilac, cylindrical, smooth, 8–11 × 3–4 microns.
Edibility　Good, sought after.
Habitat　In tufts, on stumps and trunks of various broadleaf trees, rarely on conifer trunks.
Season　Late autumn up to first frost, reappearing mid winter (if mild), then in spring and summer with rains.
Note　This species is also cultivated on poplar wood or various vegetable remains. Some people recognize two species here: *P. ostreatus*, with a white spore print, and *P. sapidus*, with a lilac-gray spore print.

114　PLEUROTUS CORNUCOPIAE

Etymology　From Latin, "horn of plenty."
Description　Cap 5–12 cm, briefly convex and funnel-shaped, whitish or light gray, pale ochreous brown or yellowish, glabrous. Gills whitish or cap-colored, but much paler, crowded, very decurrent on stipe, anastomosing to form a net. Stipe 3–8 × 0.7–1.5 cm. often ramified, nearly central to fairly eccentric, whitish, full, almost completely covered by the usually anastomosed extension of gills. Flesh white, soft, becoming rather tough. Distinctive mealy odor, sweetish flavor. Spores whitish in mass, then pale lilac-gray, elliptical, smooth, 7.5–11 × 3.5–5 microns.
Edibility　Very good when young.
Habitat　On stumps and trunks, living and dead, of broadleaf trees, in tufts.
Season　Summer and autumn.
Note　This mushroom may prove to be the same as *P. sapidus*. The whitish to grayish oyster mushroom with the lilac-gray spore print is abundant and a choice edible.

115 PANELLUS STIPTICUS

Etymology From Greek, "astringent."

Description Cap 1.5–4 cm, brownish ochre, yellowish, or cinnamon-colored, downy-floccose or pruinous then breaking up into small furfuraceous scales, elastic, kidney-shaped, sometimes funnel-shaped and lobate. Gills ochreous or cinnamon-colored, thin, straight and crowded, connected by veins. Stipe 0.5–2 × 0.2–0.3 cm, whitish, enlarged at top, pubescent, short and lateral, full. Flesh cinnamon-ochre-colored, elastic, thin, leathery in stipe. Somewhat acid odor and very astringent flavor. Spores white, elliptical, smooth, 3–6 × 2–3 microns, amyloid.

Edibility Inedible, possibly poisonous.

Habitat On trunks, stumps, and dead branches, particularly on broadleaf trees, mainly oak.

Season All year.

Note Some varieties of this fungus are characterized by the color and abundant pubescence of the cap. In North America there is a markedly luminescent variety (a fairly common feature of fungi growing on wood). *P. mitis* has a white, glabrous cap, stipe of the same color, downy-velvety, grows mainly on conifers, flesh sweetish. *P. serotinus* is rather large species, cap olive-green with fine downy hair, gills pale straw-yellow or orange, stipe yellowish, covered with small, brown, cottony scales.

116 PANUS TIGRINUS

Synonym *Lentinus tigrinus.*

Etymology From Latin, "marked like a tiger."

Description Cap 3–8 cm, thin, supple, fairly leathery, silvery white or cream-colored, variegated with small fibrillose scales, adpressed, brown or blackish, margin often split in dry weather. Gills white then yellowish, decurrent, thin and crowded. Stipe 3–5 × 0.5–1.5 cm, whitish, blackish brown at base, tough, often narrowing toward the base and rooting, thinly squamulose, with a short-lived ring at top, residual, with a white veil, often curved. Flesh white, blackish brown at base of stipe. Strong milklike odor, flavor somewhat acid. Spores white, elliptical, smooth, 8–12 × 3–3.5 microns.

Edibility Inedible because of the texture.

Habitat In frequently tufted groups, on the stumps of broadleaf trees or on buried rotting wood.

Season Late spring, summer, and autumn.

Note *Lentinus lepideus* has a cap which is never funnel-shaped; it is ochreous, with brownish, particularly coarse scales, raised on the stipe, forming a ring at the top; it has a smell of aniseed and grows on conifers.

117 LENTINELLUS COCHLEATUS

Etymology From Greek, "spiraled."
Description Cap 2.5–9 cm, initially flesh-colored, tending to turn paler, often chamois-colored, or reddish brown, usually eccentric, irregular, lobate and contorted, funnel-shaped, edge faintly spiraled, open on one side, imbricate. Gills white then cream-colored with pinkish shading, very decurrent, very crowded, small, edge serrate when mature. Stipe 3–9 × 0.5–1.5 cm, pale flesh-colored or reddish-brown, darker toward base, central or lateral, sulcate, often connate at base, normally hollow. Flesh thin, whitish turning to pinkish or reddish, tough, elastic, leathery. Pleasant odor of aniseed, sweet flavor. Spores white, globose, smooth, 5–6 microns, amyloid.
Edibility Can be eaten, but leathery.
Habitat On or at foot of stumps, mainly beech, tufted.
Season Summer and autumn.
Note *Lentinus* and *Lentinellus* are two similar genera easily recognized by their serrate gill edges. *Lentinus* has a mostly central stipe and spores that are elliptical, smooth, and non-amyloid. *Lentinellus* has an eccentric to lateral or absent stipe and spores that are nearly round, minutely spiny, and amyloid (turn blue-black in Melzer's Reagent).

---

118 LACTARIUS BLENNIUS

Etymology From Greek, "viscid."
Description Cap 4–10 cm, olive-gray to gray-green, with darker spots, concentric, flat then depressed at center, surface viscous. Gills white with slight red shading when mature, developing olive-gray spots when touched, crowded, thin, adnate. Stipe 4–6 × 1–1.5 cm, grayish or olive-gray, lighter than the cap, hollow. Flesh white, becoming slightly gray when exposed to air, fragile. Odorless, flavor acrid. Latex white, acrid. Spores pale cream-colored, elliptical, cristate, partly reticulate, 7.5–8.5 × 5.5–6 microns, amyloid.
Edibility Not recommended because of the acrid flavor.
Habitat Very frequent in broadleaf wood, especially beech, often hidden among dead leaves.
Season Summer and autumn.
Note This is a member of the *Lactarius* group with the cuticle fairly or very viscous; flesh and drops of latex take on greenish-gray or olive-colored shading when exposed to air. The most common species include: *L. pallidus,* ochre with pale pinkish shading; and *L. trivialis,* gray-brown with violet highlights. Only *L. trivialis* is known in North America, where it grows in the Pacific Northwest.

119 LACTARIUS FULVISSIMUS

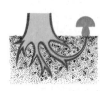

Etymology   From Latin, "deep reddish white."
Description   Cap 2.5–7 cm, at first red-brown, paling gradually with age, retaining a darker coloration at center, convex, depressed at disc, then funnel-shaped–craterlike, surface dry, finely rugose. Gills cream-colored, with pale flesh- or orange-colored shading, averagely crowded, decurrent. Stipe 3–4 × 0.9–1.6 cm, orange, reddish orange at base, short, squat. Flesh reddish white, thick, hard. Buglike odor. Latex white, constant, not plentiful, with a sweet flavor but a little harsh on the throat. Spores cream-colored, almost spherical, spiny, often with crests and connecting lines, 7.5–8.5 × 6.5–7.5 microns, amyloid.
Edibility   Can be eaten, but flavor mediocre.
Habitat   In broadleaf and coniferous woods, lowlands and highlands, especially on clayey-calcareous ground.
Season   Summer and autumn.
Note   Easily confused with very similar species: *L. ichoratus* has a brighter, more fibrillose cap; the cap of *L. mitissimus* is decidedly more orange in color; *L. subdulcis* has softer coloration; *L. rubrocintus* is much larger and has a distinctive reddish ringlike zone in upper part of stipe. None of these species is known to occur in North America.

120 LACTARIUS GLAUCESCENS

Synonym   *Lactarius piperatus glaucescens.*
Etymology   From Latin, from the blue-green color of its flesh when cut.
Description   Cap 5–15 cm, whitish, initially convex, then funnel-shaped, surface bare, not rough, at most cracked. Gills yellowish white, crowded, decurrent, almost anastomosed. Stipe 2–6 × 1.5–2.5 cm, whitish, full, may have small pits with blue-green spots. Flesh creamy white, turning blue-green when cut, hard. Odorless, with a hot, acrid flavor. Latex white, becoming greenish like the flesh, when drying out. Spores white, subspherical, slightly and incompletely reticulate, 7–8.5 × 6–7 microns, amyloid.
Edibility   Not recommended because of the acrid flavor, and it has been reported to cause poisoning.
Habitat   Grows in large numbers in broadleaf and coniferous woods.
Season   Late summer and autumn.
Note   This species is very similar to *L. piperatus* (**121**). The latex of these two fungi does, however, react differently to KOH treatment: it quickly turns bright yellow in the case of *L. glaucescens*, but does not alter in the case of *L. piperatus*.

121 LACTARIUS PIPERATUS

Etymology From Latin, "peppery," from the flavor.
Description Cap 5–13 cm, whitish, may develop ochre spots with age, initially convex, then flattening and eventually becoming funnel-shaped, surface dry, rugose near margin. Gills creamy white with pinkish shading, slightly decurrent, forked near stipe. Stipe 3–8 × 1.5–2.5 cm, same color as cap, cylindrical, irregular, full, rugose, finely pruinous. Flesh white, hard. No odor, acrid flavor. Latex white, immediately hot and acrid. Spores white, subspherical, light, imperfect reticulum, 7–8.5 × 5.5–6.5 microns, amyloid.
Edibility Not recommended because of the acrid flavor.
Habitat In broadleaf woods, rarely beneath conifers.
Season Summer and autumn.
Note This fungus, like *L. glaucescens* (**120**), has flesh and latex with an acrid flavor that burns the tongue, which is why neither can be used in cooking. But in some Eastern European countries they are eaten after a lengthy period of treatment similar to the preparation of sauerkraut.

122 LACTARIUS LILACINUS

Etymology From Latin, "lilac-colored."
Description Cap 4–7 cm, from brown-pink to pink-lilac, flat-depressed, cuticle dry, finely furfuraceous, then squamulose. Gills ochre-pink, then very like cap color, not very crowded, slightly decurrent. Stipe 4–8 × 0.6–0.8 cm, slender, enlarged at base, solid then hollow, ochre-pink. Flesh whitish lilac. Slight odor of chicory. Latex whitish, watery, turning greenish gray when drying on gills, quite sweet. Spores white, ovoid, incompletely reticulate, 7.5–8 × 6 microns, amyloid.
Edibility Not recommended, though not poisonous.
Habitat This species grows exclusively under alder, where it is quite frequent and plentiful.
Season Summer and autumn.
Note A very similar species is *L. spinosulus*, growing in the same habitat; it is smaller, more conspicuously violet in color, with a small pointed umbo at the center of the cap, which is slightly scaly. Both belong to the *Lactarius* group having a completely dry, downy-hairy cap. Neither species is known to occur in North America.

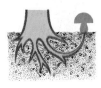

123 LACTARIUS PORNINSIS

Etymology After the French mycologist Pornin.

Description Cap 3–8 cm, orange, with brighter red or orange zonations, convex, becoming slightly depressed at center, cuticle slightly viscous as a result of humidity. Gills first ochreous cream-colored, then ochreous orange, crowded, slightly decurrent. Stipe 3–6 × 0.8–1.5 cm, orange but lighter than cap, soon becoming hollow. Flesh slightly reddish. Unpleasant smell and bitter flavor. Latex white, unchanging. Spores white, ovoid, reticulate, 7.5–9.5 × 5.5–7 microns, amyloid.

Edibility Can cause mild stomach troubles.

Habitat It grows exclusively in larch woods, where it is very frequent and plentiful.

Season Summer and autumn.

Note This mushroom is not known in North America. Many bitter or acrid species of *Lactarius* are known to cause stomach upsets. However, in Europe certain species are sold in markets and are prepared in ways that remove the irritant compounds.

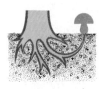

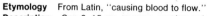

124 LACTARIUS SANGUIFLUUS

Etymology From Latin, "causing blood to flow."

Description Cap 6–15 cm, orange- or wine-red, sometimes with darker zonations, developing green spots, flattened, margin involute. Gills winy-orange, may develop some green markings, crowded, barely decurrent. Stipe 4–6 × 1.5–2.5 cm, orange with wine-colored markings, covered with whitish pruinescence, may have small wine-red pits. Flesh only faintly colored, whitish, becoming blood-red when touched. Latex wine-red, slightly bitter. Spores white, subspherical, cristate, reticulate, 8–9.5 × 6.5–8 microns, amyloid.

Edibility Excellent cooked.

Habitat In coniferous woods, especially beneath pines, and prefers warmer latitudes.

Season Autumn.

Note The very similar fungus from western North America, commonly known as *L. sanguifluus,* is now believed to be a distinct species, *L. rubrilacteus. L. sanguifluus* is certainly more tasty than the more famous *L. deliciosus,* with which it is often confused. All *Lactarius* with wine- or carrot-red latex are edible. It is a good rule to consume only this group of *Lactarius,* excluding those with latex initially white or watery; although nontoxic, they often taste acrid.

125 LACTARIUS DELICIOSUS

Common name Saffron milk cap; milk agaric.
Etymology From Latin, "delicious."
Description Cap 5–12 cm, bright orange, deeper colored zonations, may turn slightly green with age, convex, slightly depressed at center, margin involute. Gills bright orange, sometimes green when mature, crowded, slightly decurrent. Stipe 3–6 × 1–2 cm, cap-colored, with small and more brightly colored pits. Flesh whitish, orange at edge because of latex, soft. Odor fragrant, flavor slightly bitter. Latex orange, carrot-red. Spores creamy white, rounded, cristate, reticulate, 7.5–9.5 × 6.5–7 microns, amyloid.
Edibility Good.
Habitat In coniferous woods, especially pine.
Season Autumn.
Note This species belongs to the *Lactarius* group with latex that is colored from the first, and also includes: *L. indigo,* a blue mushroom with blue latex; *L. paradoxus,* a grayish blue mushroom with dark reddish brown latex; *L. subpurpureus,* a pinkish-to-reddish mushroom with reddish brown latex; *L. rubrilacteus,* an orange mushroom with reddish latex. All stain greenish in age or on bruising, and are found under conifers.

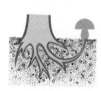

126 LACTARIUS CONTROVERSUS

Etymology From Latin, "turned the other way."
Description Cap 7.5–25 cm, whitish with concentrically placed pink spots, convex becoming funnel-shaped, often irregular, cuticle moist and viscous, usually with not very conspicuous zonations. Gills a distinctive flesh-pink color, crowded. Stipe 3–5 × 1.5–3 cm, whitish, squat, hard. Flesh white. No odor, acrid flavor. Latex white, does not change when exposed to air, flavor very acrid. Spores cream-colored, subspherical, finely reticulate, interrupted, 6–6.5 × 5 microns, amyloid.
Edibility Not recommended.
Habitat Beneath broadleaf trees, prefers poplars, in grass, under willows.
Season Summer and autumn.
Note An easily identifiable fungus because of its exceptional size and, above all, by the pink gills which single it out from the other white *Lactarius* species, from which it also differs because of the cap zonations. It is one of the largest and the most fleshy of this *Lactarius* group; the flesh of the cap is much thicker than the width of the gills.

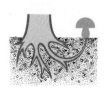

127 LACTARIUS VOLEMUS

Etymology From Latin, "filling the palm of the hand," an adjective used to describe a very large variety of pear.

Description Cap 7–15 cm, red-brown to reddish orange, convex then flat, depressed at center, surface dry, cracked in old specimens. Gills pale ochre-cream, spotted with brown when rubbed, averagely crowded, slightly decurrent. Stipe 6–12 × 1.5–2.5 cm, cap-colored, tough, full, pruinous. Flesh whitish, darkening slightly, thicker in cap than the width of gills. Strong distinctive odor of herring or Jerusalem artichoke. Latex white, very plentiful, sweet. Spores white, spherical, cristate-reticulate, 8–12 × 7–11 microns, amyloid.

Edibility Good, and popular.

Habitat Grows in coniferous and broadleaf woods, common and very plentiful when found.

Season Summer and autumn.

Note *L. luteolus,* similar, with abundant white latex and a fishy odor, has a buff cap with a whitish bloom; *L. hygrophoroides* looks like *L. volemus* from above, but its gills are very distant and it has no odor; *L. corrugis* is usually a darker brownish red with a wrinkled cap. Other similarly colored species have intensely acrid latex and probably should not be eaten!

128 LACTARIUS ACERRIMUS

Etymology From Latin, "very acrid."

Description Cap 5–13 cm, yellow-ochre, somewhat tinged with orange, zoned by darker watery markings, massive, convex, slightly depressed at disc, then conspicuously funnel-shaped. Gills from whitish to yellow-ochre, decurrent, forked and anastomosed near stipe. Stipe 3–7 × 2–3 cm, whitish with small, scattered, cap-colored markings, soft, massive, irregular, tapering at base. Flesh whitish, may turn yellow at base of stipe when cut, then turns slowly reddish and finally blackish gray. Odor fruity, flavor very acrid. Latex white, unchanging, scant hot-acrid flavor. Spores cream-colored, subspherical, cristate, 10–13 × 8–10 microns, amyloid.

Edibility Not recommended because of the flavor.

Habitat In broadleaf woods, especially oak.

Season Late summer and autumn.

Note This fungus differs from the other members of the *L. zonarius* group by its much larger spores and the two-spored basidia. *L. zonarius* has similar coloration and a very strong fruity smell; *L. insulsus* has a much more orange and zoned cap, stipe covered with conspicuous small pits; *L. zonarioides* grows only beneath conifers in mountainous regions; when touched, the gills become marked with greenish. None are known in North America.

129 LACTARIUS CILICIOIDES

Etymology From Greek, "resembling cilice," (a goat-hide fabric).

Description Cap 3–7 cm, whitish, tending to turn ochre, convex then flattened with center slightly depressed, margin very involute and covered with a feltlike layer of wooly hairs a few millimeters in length, center bare, though slightly viscid. Gills cap-colored, crowded and adnate. Stipe 2–4 × 1–1.5 cm, whitish then ochreous, with a few pits narrowing toward base. Flesh white. Slight odor of geranium, acrid. Latex white, unchanging, acrid. Spores pale cream-colored, elliptical, cristate, 6–7.5 × 4.5–5.5 microns, amyloid.

Edibility May cause very serious intestinal disorders.

Habitat In broadleaf woods, particularly beneath birch, or near birch trees in grassland.

Season Summer and autumn.

Note This can be considered a pale form of *L. torminosus* (**131**), differing in color and in the slightly smaller size of the spores. Very similar *L. pubescens* is smaller and grows almost exclusively in northern Europe.

130 LACTARIUS SCROBICULATUS

Etymology From Latin, "pock-marked" or "pitted."

Description Cap 6–20 cm, yellow or reddish yellow, faintly zoned, large, fleshy, convex, depressed at center, margin conspicuously involute, wooly with numerous not very long hairs. Gills yellowish, lighter than cap, crowded, slightly decurrent. Stipe 3.5–6 × 2–3.5 cm, whitish, with numerous yellow or yellowish red pits, hard and hollow. Whitish flesh, yellowing when cut because of the latex, tending to blacken with time, quick to rot. Odor slightly fragrant, flavor acrid to burning. Latex white, turning immediately sulfur-yellow when exposed to the air. Spores pale cream-colored, rounded, finely reticulate, 8–8.5 × 6.5–7.5 microns, amyloid.

Edibility Inedible, possibly poisonous.

Habitat In coniferous woods in mountains, occasionally beneath willows.

Season Summer and autumn.

Note This is a very beautiful *Lactarius* in form, color, and above all for the way its white latex turns immediately yellow when exposed to air. Very similar species include *L. resimus* and *L. citriolens,* with the latex also turning yellow, but without pits on the stipe. *L. representaneus* looks like *L. scrobiculatus,* but the white latex turns violet on exposure to air.

131 LACTARIUS TORMINOSUS

Etymology From Latin, "causing colic."
Description Cap 4–12 cm, varying in color from pale orange to brick-red, convex, then flattened, depressed at center; the surface is covered with dense long hairs, and has fairly conspicuous dark orange-red zones. Gills light salmon-ochre, crowded, slightly decurrent. Stipe 2–6 × 1–2 cm, whitish to orange-red, soft. Flesh whitish, thick. Slightly fruity odor. Latex white and very acrid. Spores pale cream-colored, elliptical, cristate, reticulate, 7.5–10 × 6–8 microns, amyloid.
Edibility Causes severe gastrointestinal disorders unless thoroughly cooked.
Habitat In broadleaf woods, particularly beneath or near birch.
Season Summer and autumn.
Note This is the commonest of the hairy-capped *Lactarius* species. In addition to *L. cilicoides* **(129)** and *L. pubescens*, which are much lighter in color, there is *L. mairei*, which has a reddish ochre cap with no hint of pink, a much stronger fruity smell, and grows beneath broadleaf trees with a preference for Mediterranean-type scrub with cistus and evergreen oak or ilex.

132 LACTARIUS TURPIS

Synonym *Lactarius necator; Lactarius plumbeus.*
Etymology From Latin, "foul," "disgusting."
Description Cap 8–25 cm, olive-grayish brown, greenish at edge, very dark, convex, depressed at the center, margin pubescent, surface viscous in wet weather. Gills creamy white becoming brown when rubbed, crowded. Stipe 4–8 × 2–5 cm, olive-brown, paler than the cap, hard, soon becoming hollow, slightly viscous in wet weather. Flesh whitish, slightly brown-tinged when exposed to air, becoming black when drying out, thick, hard. Virtually odorless. Latex white and acrid. Spores pale cream-colored, rounded, reticulum somewhat interrupted, 6.5–8 × 5–7 microns, amyloid.
Edibility Not recommended because of acrid flavor.
Habitat In broadleaf woods, prefers flinty ground.
Season Late summer and autumn.
Note This fungus has dark blackish coloration and one can find old specimens that have become black all over. The cap surface turns a beautiful violet in contact with ammonia. This European species has two North American look-alikes: *L. sordidus*, which has a nonviscid to slightly tacky, dirty yellowish brown cap with greenish black tones; and *L. atroviridis*, an eastern species that has a green cap and stipe with dark green spots.

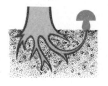

133   LACTARIUS VELLEREUS

Etymology   From Latin, ''velvety.''
Description   Cap 8–25 cm, white to ochre-tinged, at first convex then depressed at center, eventually becoming funnel-shaped, irregular, surface finely velvety. Gills white, with pale blue highlights, then reddish cream-colored, distant, fairly decurrent. Stipe 3–6 × 2–3 cm, cap-colored and likewise velvety, soft, large, short and irregular. Flesh white, slightly yellow-brown when cut, with greenish shading, hard, thick. No odor, flavor peppery and acrid. Latex white, unchanging, very acrid. Spores white, subglobose, with slight reticulum, incomplete, 9–12 × 7.5–10 microns, amyloid.
Edibility   Not recommended because of the acrid, peppery flavor.
Habitat   In broadleaf woods, especially in calcareous ground.
Season   Summer and autumn.
Note   This species differs from *L. piperatus* (**121**) and *L. glaucescens* (**120**) by having much more distant gills and the velvety cap and stipe, and like the others, it is eaten in eastern Europe after fermentation. *L. deceptivus* has a smooth cap with a floccose roll about the margin; it becomes somewhat scaly and cracked in age.

134   RUSSULA VITELLINA

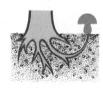

Synonym   *Russula lutea.*
Etymology   From Latin, ''yolk-yellow.''
Description   Cap 2–6.5 cm, lemon or chrome-yellow, first uniform then fading to cream-colored at the edge, globose then slightly depressed, with cuticle detachable, at first slightly viscid, shiny, with margin tuberculate-sulcate when mature. Gills eventually becoming fairly distant, sometimes with rare lamellae, forked, ochreous or orange-yellow, darkening slightly with age. Stipe 2–6 × 0.7–1.5 cm, white, quite rugose, slightly enlarged at the top, almost hollow, pithy. Flesh white, fragile. Slight smell of vinegar in old specimens, sweet flavor. Spores ochreous yellow, elliptical, with isolated pointed warts, 7.5–11 × 6.2–9 microns, amyloid.
Edibility   Can be eaten, but of little interest.
Habitat   Prefers broadleaf woods.
Season   Summer and autumn.
Note   May be confused with the yellow forms of *R. chamaeleontina*, a species varying a great deal in color. It can sometimes be identified without a microscope by the fact that the flesh emits a conspicuous roselike smell as it ages.

135 RUSSULA ADUSTA

Etymology From Latin, "burnt," from its color.

Description Cap 5–15 cm or more, whitish becoming tinged with light reddish brown tending to grayish or sepia-brown, convex-expanded becoming eventually depressed at center, cuticle adnate, shiny and smooth, viscous in damp weather, margin thick, undulate. Gills pale cream-colored or ivory, with a few pale ochre highlights, crowded, distant when mature, interspersed with numerous lamellae, arcuate, rounded at stipe, narrowing frontwards. Stipe 3–10 × 1.5–4 cm, or more, white, tinged with reddish at base, which is grayish brown, cylindrical, narrowing or enlarged in the lower part, full, hard, pruinous. Flesh creamy white turning gradually pink then soot-colored, thick, firm. Slight but distinctive odor of old casks, flavor sweet. Spores white, hemispherical-ovoid, with crowded warts, 7–10 × 6–8 microns, amyloid.

Edibility Fairly poor quality.

Habitat In sandy pinewoods, also beneath Norway spruce.

Season Late summer and autumn.

Note This belongs to the *Russula* group characterized by blackening all over with age. Another, *R. nigricans,* has large distant gills, and its flesh when cut turns red before becoming black.

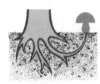

136 RUSSULA FOETENS

Etymology From Latin, "fetid-smelling."

Description Cap 6–18 cm, pale ochre at edge, yellow-brown at disc, then tawny ochre-yellow, tinged with reddish brown, at first globose then flattened at top, flat, eventually slightly depressed, cuticle fairly detachable, radially sulcate to margin, which is thin, undulate-lobate. Gills whitish then cream-colored with reddish brown markings, unequal, large, some forked, connected by veins; young specimens produce watery droplets in damp weather. Stipe 4–12 × 1–4 cm, whitish marked with brownish starting at base, tough, first solid then hollow and fragile, cylindrical. Flesh whitish turning brownish when exposed to air. Odor and taste disagreeable. Spores pale cream-colored, ovoid, with thick conical or obtuse warts, 7.5–10 × 6–9 microns, amyloid.

Edibility Inedible because of flavor.

Habitat In groups in broadleaf and coniferous woods, in wet ground, in lowlands and mountains.

Season Early summer to midautumn.

Note Not known in North America, but similar species are: *R. subfoetens,* smaller with a faint odor and taste; *R. fragrantissima,* with unpleasant odor with pleasant component smelling like marzipan; and *R. laurocerasi,* more yellow in color and more fragrant. None is palatable.

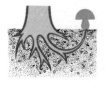

137 RUSSULA SARDONIA

Etymology From Greek, on account of acrid flavor.
Description Cap 3–10 cm, violet-purple, black at center or fading to olive-cream, often all purple-red, hemispherical then flat, depressed, cuticle adnate, slightly viscous, smooth or slightly cracked, shiny, margin thin, slightly striate. Gills pale then cream-colored, typically tinged with lemon-yellow, crowded, rigid, thin, never obtuse frontwards, arcuate then straight, producing small droplets of water in young specimens. Stipe 3–8 × 1–3 cm, violet-purple, bluish violet more marked at center, and also whitish or orange-yellow at base, solid, rigid, pruinous at top, rugose in mature specimens. Flesh whitish tinged with lemon-yellow, wine-red beneath cuticle, thick, hard, rigid, succulent. Odor of dried fruit, very acrid flavor. Spores ochreous-cream, hemispherical-ovoid, warty, reticulate, 6.5–8.5 × 6–7.5 microns, amyloid.

Edibility Inedible because of very acrid flavor.
Habitat Beneath conifers, prefers pinewoods, on sandy soil in lowland areas.
Season Late summer and fall.
Note In contact with ammonia the flesh and gills give a typical and very marked pinkish red reaction. It resembles two other species associated with conifers, that don't react to ammonia, however, and which have white or whitish gills, *R. queletii*, and *R. torulosa*. None in North America.

138 RUSSULA ATROPURPUREA

Synonym *Russula krombholzii.*
Etymology From Latin, "black and purple-red."
Description Cap 5–12 cm, dark purple-red with central part blackish, tinged with ochre-brown, hemispherical then convex, eventually slightly depressed and often undulate-lobate, cuticle detachable at edge, slightly viscid, shiny with tendency to discolor. Gills forked at base, whitish cream, arcuate, almost pointed at margin. Stipe 7–7.5 × 1–2 cm, white, unchanging or sometimes slightly tinged with ochre-brown, especially at base, rarely tinged with pink, solid, quite regular, pruinous at top. Flesh whitish or sometimes tinged with ochre-brown in stipe, purple-red beneath cuticle. Slight fruity odor, acrid flavor, especially in gills. Spores white, ovoid, reticulate, 7.5–8 × 6–7 microns, amyloid.

Edibility To be avoided, like all the acrid species.
Habitat Beneath conifers and broadleaf species, prefers oak and beech, especially on acid ground.
Season Spring and autumn.
Note The variety *depallens* usually grows in wetlands, is discolored, ochreous, with flesh tending to turn grayish with age. There are various other varieties with different colored caps and flesh and different flavors.

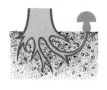

139   RUSSULA LEPIDA

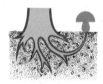

Etymology   From Latin, "pretty," from its color.

Description   Cap 3–11 cm, beautiful red color, sometimes vermilion and crimson, lighter at edge, sometimes with discolored areas, or even whitish and lemon-yellow at disc in the variety *lactea*, rounded then convex, eventually fairly flattened, fleshy, hard, margin curved, regular or slightly undulate, cuticle not detachable. Gills gypsum-white with cream-colored highlights, sometimes pinkish toward cap edge, crowded, with few lamellae, forked-anastomosed near stipe, obtuse frontwards. Stipe 3–8 × 1–3 cm, white, fairly speckled with crimson-red, especially at base, hard, fragile, full, club-shaped, sometimes short and almost cylindrical, rugose lengthwise. Flesh white, pink beneath cuticle, thick, hard, compact, granular when broken. Odor of cedar or menthol, similar flavor or resinous and somewhat bitter. Spores pale cream-colored, subglobose-ovoid, with obtuse warts, 6–8 × 6–8 microns, amyloid.

Edibility   Mediocre, must be parboiled.

Habitat   Coniferous and broadleaf woods, prefers beech.

Season   Early summer to late autumn.

Note   The flavor is sometimes remarkably minty, and cap cuticle is opaque, resembling colored gypsum. *R. rosacea* grows under western conifers and is very acrid.

---

140   RUSSULA EMETICA

Common name   Sickener.

Etymology   From Greek, "causing vomit."

Description   Cap 4–10 cm, varies greatly in color, normally bright red with no violet or pale purple or greenish coloration, but varieties may be white, pinkish, yellow, ochreous, globose then convex, eventually depressed with margin thin, briefly sulcate, with cuticle viscid, shiny, may become slightly rugose as it dries, easily detachable with red subcuticle. Gills white then slightly cream-colored, fairly distant, thin, rounded at stipe, almost free, fat toward margin. Stipe varies in length, up to 9–10 cm high in the variety *longipes*, fairly club-shaped at base, white. Flesh white, typically red beneath cuticle, fragile. Odor slightly fruity to puffball-like, acrid. Spores white, ovoid, warty, very reticulate, 7.5–12.5 × 6.2–9.2 microns, amyloid.

Edibility   Slightly poisonous.

Habitat   Typically in peat bogs in sphagnum, always close to conifers or birch.

Season   Summer and autumn.

Note   There are many varieties. The distinctive features of the species are: cuticle viscid, completely detachable, spores with conspicuous reticulum, flesh acrid, weak reaction to guaiacol, typically lives in sphagnum moss. *R. silvicola* is very similar but grows in woods, rarely in moss.

141 RUSSULA DELICA

Etymology From Latin, from the absence of any latex.
Description Cap 6–15 cm, whitish, often with ochreous brown speckling, hemispherical then convex, finally with a wide, deep depression, cuticle not detachable, dry, pruinous, with radial fibrils, rugose, sometimes cracked, margin thick, curled then straight, cap-colored. Gills whitish then pale ochre, tinged rust-colored, distant, unequal, anastomosed, forked, broad, adnate. Stipe 2–5 × 1–3 cm, white and faintly brownish, hard, full, cylindrical, sometimes truncated-conical, shiny, pruinous, may become slightly rugose with age. Flesh whitish, thick, hard. Odor of fruit or fish, flavor sweet but gills acrid. Spores creamy white, broadly ovoid, rounded, with mainly obtuse warts, 7–10 × 6–8.5 microns, amyloid.
Edibility Mediocre.
Habitat On calcareous and siliceous ground, in hills and mountains, and in broadleaf woods.
Season Late spring to late autumn.
Note *R. brevipes* is common in North America and resembles the European *R. delica* but with gills that are almost crowded instead of distant. *R. brevipes acrior* is more acrid in taste and has a bluish green tint at apex of stipe or on gills.

142 RUSSULA AERUGINEA

Etymology From Latin, "copper-colored."
Description Cap 4–10 cm, with various shades of green or steel-gray or, more especially, gray-green with center dark, and even entirely pale ochreous, with a hint of grass-green at center, often with small rust-colored spots, convex then flat and depressed, cuticle two-thirds detachable, slightly viscous, shiny when moist. Gills whitish with ochreous spots, crowded, forked at attachment, delicately intervenose, arcuate then faintly ventricose, obtuse at margin, crumbly. Stipe 5–7 × 1–2 cm, whitish, tinged with yellowish-brown at base, flared beneath gills, narrowing in lower part, full and sturdy. Flesh whitish tending to turn gray when exposed to air, thick, soft. Odor pleasant, flavor slightly peppery then sweet. Spores fairly markedly cream-colored, elongated, ovoid, with obtuse warts, 6–7.5 × 5–6 microns, amyloid.
Edibility Edible, slightly poisonous when raw.
Habitat In groups in coniferous forests, more rarely beneath aspen and birch.
Season Summer and autumn.
Note The flesh reacts conventionally to ferrous sulfate, turning pinkish quite quickly. Other green *Russulas* are dry or have a cracked cap.
Caution Slightly poisonous when raw.

143   RUSSULA OCHROLEUCA

Etymology   From Greek, "white-ochre-yellow."
Description   Cap 4–11 cm, yellow but variously tinged, pale lemon-yellow, yellowish ochre spotted with orange or brownish at center, late in season often light olive-yellow or ochreous greenish gray at center, convex-umbilicate then flat, slightly depressed, cuticle half detachable, moist and shiny, margin thick, curved, sometimes lobate. Gills pale cream-colored or faintly pale yellow with a few small brownish markings with age, averagely crowded, unequal, intervenose, ventricose, slightly obtuse frontwards. Stipe 3–7 × 1.5–2.5 cm, white, slightly grayish, spotted with brownish yellow from base upward, cylindrical, sometimes club-shaped, flared beneath gills, full, soft, slightly pithy at top. Flesh white, grayish at top of stipe, thick, soft then tough. Odor pleasant, flavor varying from piquant to sweet. Spores white, ovoid, aculeate, 7–9 × 6.3–9 microns, amyloid.

Edibility   In small quantities only, because of taste.
Habitat   In lowland and mountainous areas, on acid ground in coniferous woods.
Season   Summer and autumn.
Note   Another yellow *Russula, R. lutea,* has a clear yellow cap and flesh that doesn't discolor. Some varieties are so different from this typical variety that only a specialist can identify them.

144   RUSSULA OLIVACEA

Etymology   From Latin, "olivelike," from color of cap.
Description   Cap 8–18 cm, olive-green especially at center, speckled with ochre, brown, gray, purple- or wine-red, at first hemispherical then convex, then flat-convex, eventually flat-depressed, cuticle quite adherent, thin, toughish, dry, with concentric cracking, margin thick, recurved. Gills cream-colored then ochreous yellow, crowded then distant, forked, intervenose, often ventricose, crumbly. Stipe 4–10 × 2–3.5 cm, white tinged with pink, spotted brownish toward base, sturdy, cylindrical or various forms, slightly obese, full, rugose, reticulate, sometimes even cracked. Flesh whitish turning yellowish, brownish with age, thick, compact. Pleasant fruity odor, hazelnut flavor. Spores ochre-yellow, ovoid, with strong, long spines and obtuse warts, 8–11 × 6–10 microns, amyloid.

Edibility   Good.
Habitat   Grows beneath oak and in shady beech woods.
Season   Summer and early autumn.
Note   Identifiable by the cuticle, which is often concentrically cracked, and by the flesh, which produces a conspicuous crimson-red reaction to carbolic acid.

145 RUSSULA SORORIA

Etymology From Latin, "sister."
Description Cap 5–11 cm, dark brown and cigar-brown at center, fading to yellowish brown, hazel-brown at edge, sometimes with olive markings, subglobose to convex, then flat, conspicuously depressed or also split, connecting with hollow of stipe, cuticle detachable, elastic, viscous, pebbled, pruinous at disc, margin first curved then straight, then recurved, striate and tuberculate, translucent. Gills whitish, often with largish brown or grayish markings, crowded then distant, unequal, very intervenose, thin, very slightly obese, narrowing to the rear. Stipe 2–6 × 1–2.5 cm, white, starting from base tinged with brownish gray, and slightly flesh-colored reddish brown, cylindrical, flared beneath gills, quickly becoming hollow, pruinous at top. Flesh whitish then slightly grayish, brownish in hollow of stipe, thick, rigid. Slightly fruity odor, fairly acrid, nauseous flavor. Spores pale cream-colored, ovoid, with conical, obtuse, or truncated warts, 6.5–8.5 × 5–7 microns, amyloid.
Edibility Not recommended because of taste.
Habitat In sandy ground, beneath coniferous and broadleaf trees.
Season Summer and autumn.
Note Very similar to *R. pectinata*, margin deeply sulcate and tuberculate, cap light ochre-brown, flesh nauseous to taste, first scid then bitterish.

---

146 RUSSULA VIRESCENS

Common name Green agaric.
Etymology From Latin, "turning green," from its color.
Description Cap 5–12 cm or more, gray- or bluish green, then brownish in mature specimens, rarely entirely whitish, first globose then convex and open, slightly depressed at center, cuticle detachable near edge, tough, dry, cracked into small adnate scales, mealy, darker than background, margin thin, curved then straight, obtuse, sometimes radially sulcate. Gills creamy white with rose-cream-colored iridescence, often with reddish brown markings, crowded, unequal, forked, often anastomosed, intervenose, arcuate then ventricose, fragile. Stipe 3–9 × 1.5–4 cm, whitish, flared beneath gills, slightly narrowing at base, full then spongy, pruinous at top, slightly rugose. Flesh whitish thick, soft. Odor initially slightly fruity, becoming unpleasant, flavor sweet. Spores whitish, broadly elliptical, with fairly distant warts, 6–8.5 × 5–6.5 microns, amyloid.
Edibility Excellent.
Habitat In broadleaf woods in grass on acid ground.
Season Late spring to autumn.
Note *R. crustosa*, equally edible, is more variable in color, usually showing patches of yellow in combination with pinkish and brownish tones.

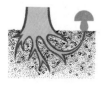

147 RUSSULA CYANOXANTHA

Etymology From Greek, "blue-yellow," from its color.

Description Cap 5–15 cm, blackish-violet, pale purple at edge and conspicuous green at disc, varying to slate-gray with lighter areas, or bluish violet or even a uniform green when mature, rounded then convex, flat, fairly depressed, cuticle two-thirds detachable, thin, viscous in damp weather, shiny, with radial fibrils and grooves, margin curved inward then obtuse, sometimes striate when mature. Gills white tinged bluish green, fairly crowded, unequal, forked, intervenose, ventricose. Stipe 5–10 × 1.5–4 cm, white, sometimes tinged lilac or reddish, with brownish markings, sturdy, even, narrowing and rooting at base, fleshy, soft then spongy, pruinous, slightly rugose. Flesh white, sometimes grayish when mature, thick, soft, moist. Odor pleasant, flavor first sweet then unpleasant. Spores white, elliptical, with small isolated warts, 7–10 × 6–7.5 microns, amyloid.

Edibility Excellent.

Habitat In broadleaf and coniferous woods.

Season Early summer to late autumn.

Note Much more common in North America is *R. variata*, also known as *R. cyanoxantha variata*, which differs by having gills dichotomously forked throughout.

---

148 RUSSULA VESCA

Etymology From Latin, "edible."

Description Cap 6–10 cm, pale flesh-pink to wine-brown, or hazel, tinged with ochre or lilac, more conspicuous or lighter olive-colored at disc, never violet, sometimes even all white in young specimens, subglobose then flat, depressed at center, cuticle hot, very detachable, with fine radial venations especially at center, margin obtuse, slightly striate. Gills whitish then cream-colored, crowded, equal, forked around stipe, intervenose, narrowing, slightly decurrent. Stipe 3–10 × 1.5–4 cm, white tending to turn yellow or develop grayish markings, brownish at base, full then spongy, slightly flared beneath gills, narrowing at base, soft. Flesh white tending to develop rust-colored or dirty yellow markings especially at base of stipe, soft, thick. Pleasant odor, tasteless or with sweet hazelnut flavor. Spores white, subglobose, with small isolated warts, 6–8 × 4.5–6 microns, amyloid.

Edibility Good.

Habitat In broadleaf and coniferous woods, usually on siliceous soil.

Season Late spring, summer, and autumn.

Note The cuticle often withdraws, revealing the flesh at cap edge. Reacts orange-pink to ferrous sulfate.

149 RUSSULA XERAMPELINA

Etymology From Greek, "color of dried vine leaves."
Description Cap 5–11 cm, purple-, crimson-, pale lilac-red with center blackish, often brownish wine-colored, with olive-ochre discoloration, convex then flat, slightly depressed when mature, cuticle not detachable, slightly viscous when damp, shiny in mature specimens, velvety, finely pebbled, margin thin and grooved in adult specimens, curved. Gills whitish cream-colored then pale ochre, distant when mature, with few lamellae, forked, intervenose, ventricose, obtuse frontwards, not very crumbly, slightly lardlike to the touch. Stipe 3–7 × 1.5–3 cm, whitish tinged with glowing red-pink, or even reddish wine-red, beneath gills developing brownish spots, cylindrical or narrowing from base to top, flared at top, soft then pithy, pruinous, venose-reticulate. Flesh whitish tending to darken when exposed to air, thick, soft. Odor of cooked crustaceans or herring, flavor sweetish. Spores deep ochre, ovoid, with obtuse or conical warts, 8–10 × 6.5–8.5 microns, amyloid.
Edibility Mediocre.
Habitat In coniferous and broadleaf woods.
Season Summer to late autumn.
Note To identify this very variable species, apart from odor, note that flesh turns brownish when exposed and has a grayish green reaction to ferrous sulfate.

---

150 RUSSULA MAIREI

Etymology After the French mycologist Maire.
Description Cap 4–8 cm, scarlet, blood-red, often paling, becoming yellow-ochre when rotten, very convex then flat, eventually slightly depressed, cuticle only detachable at margin, translucent, slightly viscous, slightly pebbled, as if velvety, even lobate at margin, which is curved, opaque, striate and tuberculate in mature specimens. Gills white or whitish, crowded then distant, unequal, rarely with lamellae, forked, intervenose, arcuate then straight, faintly obtuse frontwards. Stipe 2–6 × 1–2 cm, white, rarely with hint of red, ochre or brownish markings at base, even or enlarged halfway up, full, hard, striate and rugose. Flesh white, slightly yellowish in stipe, red beneath cuticle, thick, soft when mature. Odor of fruit or coconut, of honey when mature, flavor acrid. Spores white, ovoid, with not very crowded warts, 6–7.5 × 5.5–6.3 microns, amyloid.
Edibility Inedible, but can be used for flavoring.
Habitat In beechwoods in fairly calcareous ground, also beneath conifers on siliceous substratum.
Season Summer and autumn.
Note Not known to occur in North America.

151 CANTHARELLUS CINEREUS

Etymology From Latin, "ash-colored."
Description Cap 2–5 cm, funnel-shaped, depressed at the center, margin undulate, dark greenish brown, covered with adpressed scales. Decurrent hymenial folds, ramified and anastomosed, grayish. Stipe 3–6 × 0.4–0.7 cm, cylindrical, compressed, sinuate, hollow. Flesh grayish, thin. Odor of dried plums, sweetish flavor. Spores white, elliptical, smooth, 8–10 × 5–6 microns.
Edibility Good.
Habitat In broadleaf woods.
Season Summer and autumn.
Note This may be the same as *Craterellus cinereus*, and it may also be confused with *C. cornucopioides* (**156**), the delicious, aromatic horn of plenty. The horn of plenty is essentially two nearly identical species with faintly rugose hymenial surfaces, dark gray to brown coloration, and a vase-shaped structure: *C. cornucopioides* has a white spore print while *C. fallax* has an apricot-colored spore print.

152 CANTHARELLUS TUBAEFORMIS

Etymology From Latin, "trumpet-shaped."
Description Cap 3–7 cm, funnel-shaped, pierced at the center by a cavity that extends into the stipe, cuticle brownish yellow, slightly villose, edge recurved and margin sometimes undulate. Decurrent hymenial folds, ramified, relatively long, anastomosed, yellow to grayish sometimes with amethyst coloration. Stipe 2–8 × 0.3–0.8 cm, hollow, cylindrical, compressed, smooth, yellowish, often curved. Flesh thin, white. No particular odor, and a sweetish flavor. Spores white, ovoid, smooth, 9–12 × 7–8 microns.
Edibility Good.
Habitat In broadleaf and coniferous wood.
Season Summer and autumn.
Note The gill-like folds vary considerably in color, with some collections appearing very yellow and others deeply amethyst. The very similar *C. infundibuliformis,* found in the Pacific Northwest, is usually larger, yellow-brown in coloration, and has a yellowish spore print.

153 CANTHARELLUS AMETHYSTEUS

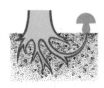

Etymology From Greek, "amethyst-colored."
Description Cap 5–10 cm, flat or depressed, covered with adherent, pale lilac-violet scales, more numerous toward margin, giving glimpses of yolk-yellow flesh, margin very curved. Hymenium with raised veins, resembling gills, anastomosed, ramified, decurrent, yellow. Stipe 3–4 × 2.5–3 cm, conical, turbinate, enlarged toward top. Flesh white then pale yellow. Fruity odor and flavor. Spores white, ovoid, smooth, 10 × 5–6 microns.
Edibility Excellent.
Habitat In beechwoods, prefers mixed with Norway spruce.
Season Summer and autumn.
Note Some consider this a variety of *C. cibarius.* Although the genus *Cantharellus* would seem to have carpophores resembling some of those formed by some gilled fungi, it is related to the Clavariaceae. The fruit-body simulates a fungus with cap and stipe, and the hymenium, which is fairly rugose, resembles gills. *Hygrophorosis aurantiaca* and *H. olida* are two cap-and-stipe species formerly included in the *Cantharellus* group. The first is entirely orange-red, with detachable, anastomosed gills; the second is pinkish cream-colored or orange with a lobate cap, sometimes very irregular, and has a strong odor.

154 CANTHARELLUS CIBARIUS

Common name Chanterelle.
Etymology From Latin, "good to eat."
Description Cap 2–10 cm, sometimes larger, convex then open and usually funnel-shaped, margin undulate, sinuate, cuticle extremely thin, transparent, yolk-yellow, glabrous. Gill-like hymenial folds, very decurrent, short, anastomosed, ramified, yolk-yellow. Stipe 3–6 × 1–2 cm, tapering from top to bottom, full, solid, cap-colored. Flesh compact, quite fibrous in stipe, pale yellow. Odor strong, often like apricots, flavor sweet. Spores pale yellow, elliptical, smooth, 7–11 × 4–6 microns.
Edibility Excellent fresh, or preserved by canning or short-cooking and freezing; drying not recommended because on rehydration they are usually tough.
Habitat Beneath coniferous and broadleaf species.
Season Late spring to late autumn.
Note There are many varieties of this fungus, which is one of the best known and most sought after. Despite its appearance it is not, systematically speaking, a gilled mushroom (agaric) because its "gills" are actually folds that develop during expansion of the cap, and are not pre-formed as in *Agaricus.* Like the French truffle it is used in liqueur-making.

155 CANTHARELLUS LUTESCENS

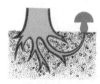

Etymology From Latin, "turning yellow."
Description Fruit-body shaped like small trumpet with upper part like a cap, 2–6 cm, consistency membranous, convex and umbilicate when immature, then open and funnel-shaped with central part closed, or connecting with hollow part of stipe, cuticle fibrillose and scaly, glabrous with age, brown or brownish gray on orange background, margin expanded, markedly lobate, undulate, curled. Hymenium from cap edge to upper third, at first smooth, then faintly sulcate with shallow venations, ramified. Stipe 5–8 × 0.5–1 cm, slender, irregular, narrowing toward base, compressed, grooved or ribbed, especially in the upper part, hollow, shiny, orange-yellow. Flesh thin, quite tough, pale yellow. Pleasant, fruity odor, somewhat of alcohol, flavor sweet. Spores white, elliptical, smooth, 10–12 × 6–7 microns.
Edibility Excellent.
Habitat Gregarious, sometimes almost tufted, in damp coniferous and broadleaf woodland in mountainous areas.
Season Summer and autumn.
Note Can easily be confused with *C. tubaeformis* (**152**), with a more raised hymenium, with gill-like folds, ramified, fairly yellow but normally pale lilac-gray when mature.

---

156 CRATERELLUS CORNUCOPIOIDES

Common name Horn of plenty.
Etymology From Latin, "like a horn of plenty."
Description Cap 2–8 cm, initially almost tubular then open and trumpet-shaped, with a thin edge, rounded then flared with margin undulate and lobate, pierced at center and hollow down to stipe, cuticle dark brown or slate-black, covered with small brown adpressed scales. Hymenium with small ash-gray wrinkles. Stipe no more than 1.5 cm wide, narrowing from top to bottom, tubulose, fibrillose, blackish. Flesh thin, blackish-gray, fairly elastic. Odor aromatic, flavor slightly astringent. Spores white, elliptical, smooth, 10–15 × 6–9 microns.
Edibility Excellent, much sought after.
Habitat In the thin litter beneath conifers and broadleaf trees.
Season Summer and autumn.
Note When dried and reduced to powder this is an excellent flavoring in soups, sauces, for roast meats, etc., and is known in Italy as "the poor man's truffle." Identical *C. fallax* is more common in eastern North America, and differs by an apricot-colored flush seen in the mature outer surface and its apricot-colored spore print.

157 LEPTOGLOSSUM MUSCIGENUM

Synonym *Cantharellus muscigenus.*
Etymology From Latin, "born from moss."
Description Fruit-body fan- or spatula-shaped, 1–2.5 cm, slate-gray when wet, ash-colored or whitish and zoned when dry, membranous, elastic, slightly undulate when fully mature. Hymenium with ramified venations, broad, starting from stipe, same color as fruit-body, anastomosed toward margin. Stipe 2–4 × 2–4 mm, small, lateral, horizontal and continuous with fertile part, villose at base. Flesh thin, grayish. No particular odor or flavor. Spores white, ovoid, smooth, 7–9 × 4–6 microns.
Edibility Of no interest because of size.
Habitat On moss.
Season All year round during wet periods.
Note This is a small, insignificant fungus with a specific habitat which it shares with other species in the genus: *L. glaucum,* which may grow by brackish water, has nonanastomosed hymenial venations; *L. lobatum,* growing in peat bogs at high altitude, with spores 8–10 × 6–7 microns, is larger (up to 6 cm) with bucklelike hyphae, also grows on *Carex. L. retirugum* is cup-shaped with a superior hymenium, and grows in twigs in moss; *L. bryophilum,* funnel-shaped, is completely white.

158 VOLVARIELLA BOMBYCINA

Synonym *Volvaria bombycina.*
Etymology From Latin, "silken."
Description Cap 5–20 cm, ochre-white toward disc, silvery white toward outside, at first parabolic then convex or almost flat, sometimes slightly umbonate, cuticle adnate, shiny, with fibrils or fibrous scales, silky, darker, margin rounded and curved then straight and regular. Gills white then pink, eventually brownish, crowded, free, with numerous lamellae. Stipe 7–10 × 0.6–2 cm, satiny white, pale yellow in lower part, easily detachable from cap, or heterogeneous, narrowing from bottom to top, sometimes subbulbous, slender, full, smooth, slightly fibrous in adult specimens. Volva whitish or ochre-white, outer cuticle often with yellowish-brown or brownish markings, persistent, free, wide, large and membranous. Flesh pure white, thin at edge, not hygrophanous, tender. Strong odor of radish or wood, flavor pleasant. Spores pale pink, elliptical, smooth, 7–9 × 5–6 microns.
Edibility Mediocre to good.
Habitat Usually solitary, on rotting broadleaf tree trunks, in holes in living trees, sometimes some yards above ground, and on hollow stumps.
Season Late spring to late autumn.
Note A very beautiful fungus, which was cultivated in northern Italy in the 19th century.

159　VOLVARIELLA VOLVACEA

Synonym　*Volvaria volvacea.*
Etymology　From Latin, "with a volva."
Description　Cap 4–10 cm, ovoid then campanulate-conical, eventually flattened, first blackish, which persists in mature fungi at disc, then with dark brown or blackish brown bars on a whitish or silvery white background, velvety. Gills long, white then pinkish, free. Stipe 5–12 × 1 cm, white, hairy, hollow when mature, volva membranous, brown or dark brownish gray externally (at least in part), whitish background, uniformly felt-like. Flesh white, soft, slightly fibrous in stipe. Very slightly earthy odor, insignificant flavor. Spores pink, elliptical, smooth, 7–9 × 5–6 microns.

Edibility　Good when young.
Habitat　In loam in greenhouses or gardens, on sawdust or other vegetable residue, also in areas where field mushrooms are cultivated.
Season　Spring to autumn; with heat, all year.
Note　Easy to cultivate. *V. speciosa* and *V. speciosa gloioce-phala* grow in grassy areas or plowed fields, and have a viscid cap (whitish in the former and sooty gray in the latter). *V. sur-recta* (*V. loveiana*) is easily identifiable because it forms small fruit-bodies on other fungi, especially on *Clitocybe nebularis*.

160　PLUTEUS CERVINUS

Etymology　From Latin, "deerlike," from its color.
Description　Cap 6–15 cm, brown to sooty brown, convex-flat, often umbonate, viscid in damp weather, fibrillose when mature, sometimes split radially, fragile, glabrous. Gills whitish then pinkish, crowded, broad, free. Stipe 5–10 × 0.7–1.5 cm, whitish with brownish, sometimes raised fibrils, heterogeneous, full, rigid. Flesh white, tender, fragile. Slight rootlike odor, flavor mild. Spores pinkish salmon, ovoid, smooth, 6–8 × 5–6 microns.

Edibility　Mediocre. See **caution.**
Habitat　On decomposing coniferous and broadleaf wood, also on sawdust, where it grows larger.
Season　Summer and autumn.
Note　*P. atromarginatus* closely resembles the above species although much more fibrillose, and darker on cap and stipe, especially when young. It differs markedly by having floccose and blackish gill edges. *P. salicinus* has white stipe with base tinged with bluish green, cap with grayish radial fibrils, and disc with small raised scales.
Caution　Don't confuse with *Entoloma*, typically terrestrial, with attached gills and angular spores! Many of these species are poisonous.

161 PLUTEUS AURANTIORUGOSUS

Synonym *Pluteus coccineus.*
Etymology From Latin, "golden yellow and wrinkled."
Description Cap 2–6 cm, first bright red then bright orange-red, deeper-colored at the disc, golden toward the margin, edge thinly striate in mature specimens. Gills white then pink, free. Stipe 4–6 × 0.5–0.8 cm, faintly narrowing toward top, often curved, white, often tinged with orange toward base. Flesh thin, whitish, orange beneath cuticle and in stipe. No particular odor or flavor. Spores pink, roundish or subglobose, smooth, 5–7 × 4–5 microns.
Edibility Can be eaten, but of little value.
Habitat Isolated or in small groups on trunks, stumps, or dead branches of broadleaf trees, prefers alder.
Season Summer and autumn.
Note Easily identified by the bright cap color. *P. leoninus* also grows on broadleaf trees, cap bright yellow, finely pruinous or velvety or bristly with yellow hairs, white stipe with yellow base, and gill edges also yellow. *P. lutescens* has a markedly yellow stipe, while the smooth cap, which is at most rugose, is brown or brownish ochre.

162 CLITOPILUS PRUNULUS

Common name Sweetbread mushroom.
Etymology From Latin, "small plum."
Description Cap 3–11 cm, white or pale yellow, or more rarely light gray, convex then flat, eventually depressed, undulate, margin involute, often lobate, thin, mealy, cuticle smooth, barely viscid if wet. Gills white then pinkish, very decurrent, narrowing at both ends. Stipe 2–6 × 1–1.5 cm, white, full, tapered toward base, pruinous or pubescent, villose at the base. Flesh white, fragile. Strong mealy odor and flavor. Spores pink, elliptical with longitudinal ridges, angular in endview, 9–13 × 5–7 microns.
Edibility Good, but see **note**.
Habitat In woodland and grassland, often under juniper.
Season Summer and autumn.
Note May be confused with certain poisonous white species of *Clitocybe*, but they have a whitish spore print. Certain toxic *Entoloma* species, which have a pinkish spore print, do not typically have long decurrent gills.
Caution When in doubt, do not eat.

163 ENTOLOMA SINUATUM

Synonym *Entoloma lividum.*
Etymology From Latin, "sinuate," from gill attachment.
Description Cap 6–20 cm, varying in color and texture, white, grayish with ochre shades, brown, pinkish, ash-colored, innate fibrils, fairly conspicuous, fleshy, not hygrophanous, convex or globose-campanulate, then flattened, barely depressed, slightly humped, dry and glabrous. Gills initially pale yellow then salmon-colored starting from stipe, retaining original color for some time at edge, fairly crowded, sinuate. Stipe 7–13 × 1.5–3.5 cm, sturdy, full, rarely slightly hollow when old, sometimes faintly curved or enlarged at base, whitish, with some yellow marking. Flesh thick, soft, slightly fibrous in stipe. Strong odor of meal turning to rotting walnut when older, sweetish, mealy flavor. Spores pink, polyhedral, 8–11 × 7–8 microns.

Edibility Very poisonous.
Habitat In groups, in sparse, dry woodland beneath broadleaf trees, especially oak.
Season Early summer to late autumn.
Note This large fleshy *Entoloma* is often eaten by mistake by people who think it is a *Pluteus,* which has free gills and grows on wood. It causes severe gastric upset.

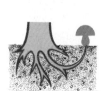

164 ENTOLOMA NIDOROSUM

Etymology From Latin, "smoked," from color of cap.
Description Cap 3–7 cm, dirty white and shiny in dry weather, ochre-ash-colored, slightly striate at margin, transparent when wet, convex then expanded, eventually concave, cuticle split radially. Gills whitish then pinkish, adnate or semifree. Stipe 3.5–9 × 0.3–1 cm, whitish or light gray, fragile, striate-fibrillose, soon becoming hollow, cylindrical, the top pruinous. Flesh white, fragile. Strong nitrous odor, fairly strong, mealy, unpleasant flavor. Spores pinkish, polyhedral, 7–10 × 6–7.5 microns.

Edibility Poisonous.
Habitat Gregarious, in broadleaf and coniferous woods, in heathland, and in scrub.
Season Summer and autumn.
Note There are many species of *Entoloma* with a mealy smell, and many of them are poisonous. In spring poisonous *E. clypeatum* is found in tufted groups beneath plants belonging to the Rosaceae family (apple, pear, plum, etc.); its cap has a conspicuous umbo like an ancient shield. Although some species of *Entoloma* are edible, these fungi are difficult to identify and differentiate from similar poisonous species; therefore, **no species of *Entoloma* should be eaten.**

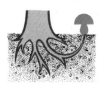

165   ENTOLOMA INCANUM

Synonym *Leptonia incana.*
Etymology From Latin, "turning white with age."
Description Cap 2–3 cm, olive-brown to grass-green, sometimes variegated with brown fibrils or small scales, thin, fragile, convex then expanded, umbilicate, slightly striate, tending to become silky and grayish with age. Gills whitish or tinged with pale yellow-green, then pinkish, fairly distant, adnexed then detached from stipe. Stipe 2.5–5 × 0.2–0.4 cm, smooth, fistular, fragile, green, greenish yellow, turns blue-green toward the base when touched, as does the surrounding white mycelium. Flesh thin, greenish. Strong unpleasant odor of mice, flavor disagreeable. Spores pink, polyhedral, 8–14 × 7–9 microns.

Edibility Of no interest because of flavor.
Habitat In fields and meadows with low grass, or in open but damp parts of woods.
Season Summer and autumn.
Note Those species with slender, tapering, smooth stipes, thin, often umbilicate caps, and gills not entirely gray or brown belong to the subgenus *Leptonia*, and are of no interest gastronomically. Several related species, such as *E. serrulatum*, have bluish caps and stipes and blue-gray gills, pinkish at maturity.

166   RHODOTUS PALMATUS

Etymology From Latin, "handlike," from its shape.
Description Cap 5–12 cm, apricot-pink, orange-hazel, flesh-pink, convex then flattened, horizontal, fairly eccentric relative to stipe, often cracked-reticulate, especially at margin which remains involute for a long period, covered by a thick and diaphanous gelatinous cuticle, emitting small, clear, orange droplets, very astringent. Gills pinkish, crowded, broad, soft, almost gelatinous, sinuate, connected by veins. Stipe 3–7 × 1–1.5 cm, whitish then orange-brown, cap-colored, fibrillose-striate, pruinous, full, eccentric or lateral. Pleasant odor but bitter, acid-astringent flavor. Spores pinkish, salmon-colored (in the spore print on paper, if the gills touch the paper, spores become dark ochre or rust-colored), subspherical, spinose, 5–7 microns.
Edibility Nontoxic but very bitter.
Habitat Normally tufted with imbricate caps on trunks, especially elm, also on posts and stakes.
Season Late summer to early winter.
Note A distinctive and unmistakable species, not very common, with caps overlapping and often closely attached because of the viscous cuticle covering them.

167 TERMITOMYCES LETESTUI

Description Cap 9–25 cm, fleshy, campanulate then convex, eventually fairly expanded, with a conspicuous cylindrical umbo that is dark brown from the small brown scales about disc. Cuticle dry and whitish, fairly pale pinkish-gray, squamulose and cracked except toward margin, which is usually appendiculate. Gills white, almost free, crowded, unequal. Stipe 12–18 × 2–3 cm or more, solid, fusiform, deeply rooting, whitish, pubescent beneath membranous ring, which is large, double, variable but complete and persistent, pendant or sheathlike. Flesh whitish, soft. Odor faint, flavor insignificant, sometimes slightly bitter. Spores white, elliptical, smooth, 6–9 × 3–5 microns.

Edibility Good.

Habitat On termite nests in tropical Africa.

Season All year.

Note The gill fungi growing on termite nests belong to the genera *Podabrella* (pink-spored) and *Termitomyces* (white-spored). The fruit-bodies form inside the tunnels and bore through the very hard layer of inert matter, forcing their way through it with a special umbo, sometimes quite pointed, called a "driller."

---

168 INOCYBE PATOUILLARDI

Etymology After the French mycologist Patouillard.

Description Cap 3–7 cm, white to brownish, passing through yellow and reddish, conical-campanulate then expanded and umbonate, cuticle dry, silky, with radial fibrils, margin split. Gills initially whitish or pinkish, then olive-brown or rusty-brown, adnexed or semifree, crowded and ventricose, edge irregular, white. Stipe 4–7 × 1–1.5 cm or more, variable, often cylindrical, sturdy, enlarged at base; bulb sometimes faintly marginate, solid, consistent, pruinous at top, fibrillose, with velar remains, white with fibrils that redden with age, turning red then brown if touched. Flesh white, faintly reddening at base. Odor quite fruity and flavor sweetish. Spores rust-ochre, elliptical, smooth, 10–13 × 5–7 microns.

Edibility Poisonous.

Habitat Often gregarious, on calcareous ground, preferably in open beechwoods or beneath linden trees, never beneath conifers. Rare in North America.

Season Late spring and throughout the summer.

Note The carpophores can be recognized by the fact that they turn from their initial pale colors to red. This slow an spontaneous color change is accelerated when the fungus is rubbed with the hands. Ingestion causes muscarine poisoning.

169 INOCYBE SPLENDENS

Etymology From Latin, "shining."
Description Cap 3–5 cm, first conical with edge raised, then convex with umbo, covered with radial fibrils, adpressed, orange-brown or pale lilac-brown, pale purple, split toward edge, joined beneath at disc, flesh beneath cuticle pale yellow, margin soon incised. Gills whitish then brownish ochreous, margin paler, adnate, sometimes slightly decurrent. Stipe 7–11 × 3–3.5 cm, sturdy, cylindrical or suddenly enlarged into an almost marginate bulb, pure white, finely striate or furfuraceous at top, full. Flesh compact, fibrous in stipe, white. Slight earthy odor and slightly nauseous flavor. Spores ochreous brown, elliptical, smooth, 9.5–11.2 × 5.5–6.2 microns.
Edibility Not certain.
Habitat Gregarious and often almost tufted, on humus-rich ground or in grassland at the edge of woods.
Season Summer and autumn.
Note A rare species showing clear distinctive features under the microscope, but not always identifiable in the wild. Most of the *Inocybe* species, unless with specific organoleptic or morphological features, are difficult to identify with the naked eye.

---

170 INOCYBE GEOPHYLLA

Etymology From Greek, "earth-colored gills."
Description Cap 1–3.5 cm, white, sometimes pale yellow in old specimens, conical then convex, flat with smallish pointed umbo, silky then fibrillose, often split or cracked at the margin. Gills whitish, then dirty gray or brownish, semifree, crowded. Stipe 4–8 × 0.2–0.6 cm, white, cylindrical, slightly enlarged at base, often supple, satiny. Veil cobwebby, fugacious. Flesh white. Spermatic odor, flavor slightly acrid. Spores brownish, elliptical, smooth, 7–10 × 4–6 microns.
Edibility Poisonous.
Habitat Gregarious in coniferous and broadleaf woods, also beneath bushes.
Season All year in warmer regions.
Note Young specimens have an ogival cap connected with the stipe by a cobwebby veil. Often found growing with it is a pinkish lilac-capped variety, *lilacina*. A distinctive western species, *I. pudica*, is white but readily bruises reddish orange.

171 INOCYBE FASTIGIATA

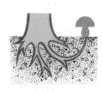

Etymology From Latin, "inclined," because of the shape of the cap.

Description Cap 2–7 cm, pale straw- or ochreous yellow, sometimes darker at the disc, conical then raised at the edge with distinct umbo; surface dry, very fibrillose, margin soon split. Gills grayish, tinged with olive, then brownish, crowded, narrow, adnexed, with lighter edges. Stipe 3–8 × 0.4–1 cm, whitish or light ochre, cylindrical or enlarged at the base, never bulbous, tapering, fibrillose. Flesh whitish, fibrous in the stipe. Spermatic odor, no flavor. Spores brownish, elliptical, smooth, 7–10 × 4–5 microns.

Edibility Poisonous.

Habitat In woods and nearby grassland.

Season Summer and autumn.

Note One variety, *I. superba,* has a cap covered with abundant silky, white, silvery fibrils. *I. obsoleta* has a faint mealy odor and grayish gills. *I. cookei,* another distinctive member of the group, has a small but conspicuous bulb at the base. Most species of *Inocybe* cause muscarine poisoning.

172 CORTINARIUS COTONEUS

Etymology From Greek, "the color of wild olive."

Description Cap 3–8 cm, olive-brown or olive-yellow tending to darken at disc, hemispherical becoming flattened, typically velvety and tomentose with fine adpressed scales, dry. Gills more rusty olive-yellow, slightly distant, adnexed, broad, with lighter and slightly denticulate edge. Stipe 4–8 × 1–3 cm, pale olive-yellow, darkening toward base until becoming almost cap-colored, characterized almost invariably by a ring-shaped mark, club-shaped and enlarged at base. Veil olive-yellow, cobwebby, fugacious. Flesh pale olive-yellow, darker at base of stipe. Distinctly radishlike odor and flavor. Spores rusty brown, almost globose or slightly elliptical, coarsely warty, 7–9 × 6.5–7.5 microns.

Edibility Suspect.

Habitat In broadleaf woods, particularly under beech.

Season Late summer and autumn.

Note The *Cortinarius* group is large, so its division into clearly defined subgenera is important. *C. cotoneus* is part of the subgenus *Leprocybe,* which includes fungi that are neither viscid nor hygrophanous, with nonsmooth caps (cuticles have a specific microscopic structure, usually red, orange, yellow or olive-colored).

173 CORTINARIUS ORICHALCEUS

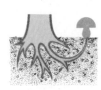

Etymology From Greek, "copper-colored."

Description Cap 4–10 cm, coppery red or tawny brick-colored, darker at disc with a greenish or blue-green area at margin, eventually entirely reddish tawny, convex then flattened, fleshy, smooth, viscid in damp weather. Gills greenish yellow then rusty olive, crowded, broad, emarginate, with undulate edge. Stipe 4–8 × 1.5–2 cm, greenish yellow, fibrillose, ending at base with a reddish-tinged marginate bulb. Veil greenish lemon-yellow, cobwebby. Flesh whitish, tinged with greenish yellow toward margin. No odor, sweet flavor. Spores rust-brown, elliptical, warty, 11–13 × 6.5–7 microns.

Edibility Not certain.

Habitat In coniferous woods in mountains.

Season Late summer and autumn.

Note All *Cortinarius* have warty spores, rusty brown in mass, finally hiding original color of gills; they also have a cobwebby veil, hence the name (*cortina*, "curtain," "veil"); the remains, colored by spores, are usually visible on stipe when mature. *C. orichalceus* is part of the subgenus *Phlegmacium*, with a smooth and viscid cap in wet weather and a dry stipe.

174 CORTINARIUS BULLIARDI

Etymology After the French mycologist Bulliard.

Description Cap 4–8 cm, red-brown in damp weather and light clay-brown somewhat tinged with reddish in dry, convex, slightly humped, smooth. Gills pale amethyst or purple then rusty brown, slightly distant, broad, sinuate-adnate, with edge denticulate and whitish. Stipe 6–8 × 1–2 cm, pale lilac-white at top, reddish below, becoming a magnificent fiery red, cinnabar-red at base, which is enlarged. Veil whitish, cobwebby, fugacious. Flesh pale lilac-whitish, then reddish, cinnabar-red at base of stipe. No particular odor or flavor. Spores rusty brown, elliptical, warty, 8.5–10.5 × 5–6 microns.

Edibility Suspect.

Habitat In dense, shady broadleaf woods, especially beneath beech, generally gregarious.

Season Late summer and autumn.

Note Part of the subgenus *Telamonia* (which includes fungi that are hygrophanous in damp weather), it is distinguished from the other flame-red–based members of this group by its larger size and by the presence of pale lilac markings on gills, at top of stipe, and in the flesh of young specimens.

175 CORTINARIUS SUBPURPURASCENS

Description Cap 5–7 cm, at first lead-colored with violet shades at margin, then brownish ochre tending to darken when rubbed, convex then flattened, fibrillose, smooth, viscid in wet weather. Gills pale violet then rust-colored, turning purple-violet if bruised, slightly crowded, emarginate-adnate. Stipe 5–7 × 1–1.5 cm, pale lilac-violet darkening toward base, with a strong tendency to turn purple-violet when rubbed, fibrillose, not very fleshy, with a not very conspicuous marginate bulb. Veil pale lilac, cobwebby, fugacious. Flesh whitish pale-lilac, unchanging. Specific not unpleasant odor and sweet flavor. Spores rusty brown, elliptical, warty, 8–9 × 4.5–5.5 microns.
Edibility Inedible.
Habitat In broadleaf woods.
Season Late summer and autumn.
Note C. purpurascens and its varieties can be identified by their fuller colors, their more massive appearance, and the tendency of the flesh to turn purple-violet. This latter feature also singles out C. porphyropus, which is smaller, lighter in color, and has no marginate basal bulb. These mushrooms all belong to subgenus *Phlegmacium*.

176 CORTINARIUS VIOLACEUS

Etymology From Latin, "violet-colored."
Description Cap 6–15 cm, dark violet sometimes tinged with pale purple-violet, hemispherical then convex-flattened, entirely velvety-tomentose, dry and fleshy. Gills dark violet then cinnamon-brown, sinuate-adnate, distant, broad, often joined at base by veins. Stipe 6–12 × 1.5–2 cm, cap-colored but slightly lighter, at first tomentose-velvety then just fibrillose, club-shaped, full then hollow. Veil violet, cobwebby, fugacious. Flesh violet, soft, spongy. Often a distinctive odor of cedarwood, flavor sweet. Spores rusty brown, elliptical, warty, 11–14 × 7–9 microns.
Edibility Can be eaten but not tasty.
Habitat In damp broadleaf woods, prefers oak and chestnut, sometimes in coniferous woods in mountains.
Season Summer and autumn.
Note An absolutely unmistakable species because of its uniform color and the conspicuously velvety appearance of the cap cuticle; identification is also helped by the odor. Some authors distinguish a C. violaceus, growing beneath conifers, with elongated gills, from a C. hercynicus, growing beneath conifers, with subglobose spores. These are the only species of subgenus *Cortinarius*.

177 CORTINARIUS MULTIFORMIS

Etymology From Latin, "changing in appearance."
Description Cap 4–10 cm, tawny-ochre, brown-ochre, tending to darken with age, hemispherical then convex-flattened, smooth, fibrillose with whitish velar remains, not striate, viscid in damp weather. Gills initially whitish then clayey and eventually rust-colored, crowded, emarginate, edge undulate and slightly serrate. Stipe 6–9 × 1.5–2 cm, whitish with tendency to turn ochreous toward base, full, with a bulb of varying shape, generally globose or almost marginate. Veil whitish, cobwebby, fugacious. Flesh whitish. Odor initially faint then slightly honeylike, flavor sweet. Spores rusty brown, elliptical, finely warty, almost smooth, 10–11.5 × 5–6.5 microns.
Edibility Can be eaten.
Habitat In mountains under conifers.
Season Late summer and autumn.
Note This common *Cortinarius* of subgenus *Phlegmacium* is not always easy to recognize because of its variable shape and the large number of similar species.

178 CORTINARIUS SPECIOSISSIMUS

Etymology From Latin, "most beautiful-looking."
Description Cap 2–8 cm, reddish brown or reddish tawny, immature specimens sometimes yellowish at margin because of velar remains, initially conical then flattened with an evident pointed umbo, finely feltlike, dry. Gills first cap-colored then rusty-brown, distant, sometimes venose on the surfaces, broad and adnexed. Stipe 5–11 × 0.6–1 cm, reddish tawny, slightly lighter than the cap, cylindrical or enlarged at the base, almost invariably with velar remains in the form of a yellowish band, solid then hollow. Veil yellowish, cobwebby, fugacious. Flesh reddish ochre. Odor mushroomy or slightly radishlike, flavor sweet. Spores rusty brown, ovoid, finely warty, 9–12 × 6.5–8.5 microns.
Edibility Lethally poisonous.
Habitat In coniferous woods, especially beneath Norway spruce, in moss or among bilberries.
Season Late summer and autumn.
Note This seasonally common fungus is as lethal as *C. orellanus* (**186**), from which it is distinguishable by the yellowish band around the stipe, and the distinctive umbo. Although neither is found in North America, the similar and lethal *C. gentilis* does occur. All are members of subgenus *Leprocybe*, the most dangerous in the genus *Cortinarius*.

179 CORTINARIUS ALBOVIOLACEUS

Etymology From Latin, "white and violet."
Description Cap 3–9 cm, pale lilac-white turning paler, fibrillose, silky, convex then flattened with a central mamelon, opaque, dry. Gills light violet-blue, then clayey-violet, eventually rusty, slightly distant, broad, emarginate, edge denticulate. Stipe 5–10 × 1–2 cm, pale lilac-white then whitish, club-shaped, also almost cylindrical. Veil whitish, cobwebby, fugacious. Flesh at first pale violet-blue then whitish lilac, soft. No odor or flavor. Spores rusty brown, elliptical, warty, 8–9.5 × 5–6 microns.
Edibility Edible but not usually pleasant-tasting, usually earthy.
Habitat In broadleaf woods, especially beech and oak, often gregarious.
Season Late summer and autumn.
Note This is a typical member of the subgenus *Sericeocybe*, which have generally smooth and dry to moist but not slimy caps, and never have red, yellow, olive, or orange coloration.

180 CORTINARIUS TRAGANUS

Etymology From Greek, referring to its goatlike odor.
Description Cap 4–12 cm, violet-amethyst tending to turn ochreous from the disc, hemispherical then convex, fleshy, first slightly squamulose then smooth, dry. Gills saffron-ochre when immature, then cinnamon-rust, emarginate, thin, slightly distant, sometimes denticulate. Stipe 6–9 × 1–2.5 cm, cap-colored, sometimes with ochreous bands, then brownish ochre, downy at base, club-shaped and bulbous, fleshy. Veil pale lilac-violet, cobwebby, fugacious. Flesh yellowish in cap, ochreous-yellow in stipe. Strong, penetrating smell of billygoat or acetylene, sometimes fruity, flavor sweet. Spores rusty brown, elliptical, warty, 8–10 × 5–6 microns.
Edibility Inedible, but may be used for flavoring.
Habitat In coniferous or broadleaf woods, isolated or gregarious.
Season Late summer and autumn.
Note Like *C. alboviolaceus*, this mushroom is a purplish member of subgenus *Sericeocybe*. It is distinguished by its ochre-cinnamon flesh, especially in the stipe, and its goatlike odor.

181 CORTINARIUS GLAUCOPUS

Etymology From Greek, "with light blue leg."
Description Cap 5–10 cm, tawny-ochreous tinged with olive, particularly at edge, hemispherical then flattened, soft, conspicuously fibrillose, viscid in damp weather. Gills bluish pale lilac, then tinged with a clayey coloration, eventually turning light cinnamon-colored, crowded, thin, emarginate, edge undulate or slightly denticulate. Stipe 4–8 × 1–2 cm, bluish pale lilac, then ochreous from base upward, fibrillose, with a conspicuous marginate bulb usually narrowing after margin. Veil whitish pale lilac, cobwebby, fugacious. Flesh whitish ochre in cap, pale lilac in stipe, and more intensely ochre at base. Slightly mealy odor, sweet flavor. Spores rusty brown, elliptical, with small wart, 7–9 × 4.5–5.5 microns.

Edibility Can be eaten.
Habitat In coniferous woods up to high altitudes, common, gregarious, often in large groups.
Season Late summer and autumn.
Note Unlike subgenus *Myxacium,* which has a slimy cap and stipe, subgenus *Phlegmacium* has a slimy to sticky cap (shiny when dry) and dry stipe; its species are further divided into those with a conspicuous marginate bulb, and those with a club-shaped to equal stipe.

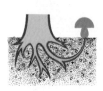

182 CORTINARIUS PRAESTANS

Etymology From Latin, "prominent" or "outstanding."
Description Cap 7–20 cm, reddish or wine-brown, margin sometimes violet, covered by a whitish violet veil which may leave scaly remains of the same color, hemispherical then convex, soft and fleshy, striate at margin in mature specimens, viscid in damp weather. Gills whitish gray tinged with pale lilac, then rust-colored, crowded, adnexed, edge slightly eroded. Stipe 10–15 × 3–6 cm, pale lilac-white, turning paler, bluish white, band-shaped velar remains on adults, club-shaped and bulbous, fibrillose, sturdy. Veil pale blue-white, cobwebby, fugacious. Flesh whitish in cap and bulb, pale blue in stipe. No odor or flavor. Spores rusty brown, elliptical, warty, very large, 12–17.5 × 8–9 microns.

Edibility Very good.
Habitat In broadleaf woods on calcareous soil, rare.
Season Late summer and autumn.
Note *Cortinarius* is not a genus much sought after for eating despite their relative abundance and the many different species (about 800 in North America, over 1000 in Europe). Unlike *C. praestans,* few are particularly tasty; some, lethal; many, mildly toxic; most, unknown. Unfortunately, this *Phlegmacium* is not known to occur in North America.

183　CORTINARIUS TRIVIALIS

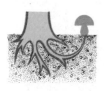

Etymology From Latin, "common."
Description Cap 3–8 cm, olive-ochre-yellow or tawny-brown, turning from campanulate-convex to flattened, often with an obtuse umbo, smooth, glutinous in damp weather, edge involute for a long period. Gills clay-colored or pale amethyst, then rust-cinnamon, not very crowded, thin, broad, adnate or decurrent with a small tooth. Stipe 4–10 × 1–1.5 cm, white at top, brownish ochre lower down, slender, tall, usually tapered and almost truncated at base, covered in wet weather by a glutinous universal veil which, as it dries, leaves behind numerous scaly bands. Veil glutinous, whitish, cobwebby, fugacious. Flesh whitish ochre, dark ochre at base of stipe. No odor, flavor sweet. Spores rusty brown, elliptical, warty, 10–15 × 6–8 microns.
Edibility Can be eaten.
Habitat In broadleaf woods, gregarious, very common.
Season Summer and autumn.
Note Typically, members of the subgenus *Myxacium*, to which *C. trivialis* belongs, are entirely viscid in wet weather. Although this species is not found in North America, two similar ones are: *C. collinitus*, with zones of tissue developing on stipe, growing under aspens; and *C. mucosus*, without zones, growing under conifers.

184　CORTINARIUS PSEUDOSALOR

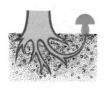

Description Cap 4–7 cm, varying from yellow-ochre to tawny-olive, almost hemispherical then flattened-convex, sometimes campanulate, with or without umbo, very viscid in damp weather; margin smooth or slightly rugose, also raised in mature specimens. Gills creamy ochre then tinged with rust color, broad, averagely crowded, adnate, with edge paler and uneven. Stipe 6–10 × 1–1.5 cm, pale blue-violet, turning paler, often whitish when mature, slender, almost cylindrical or narrowing at base which may have ochreous floccose zones; striate at apex, viscous in wet weather. Veil glutinous, whitish, cobwebby, fugacious. Flesh whitish with ochreous shading. Distinctive, slightly honeylike odor, flavor sweet. Spores rusty brown, elliptical, warty, 11–15 × 7–8 microns.
Edibility Uncertain.
Habitat In broadleaf woods, especially in lowland areas, more rarely with conifers, very common.
Season Summer and autumn.
Note *C. elatior*, also a *Myxacium*, is similar in odor and coloring, with stipe sometimes pale blue, but it differs in the cap, which is deeply striate at margin; the gills, which are venose; and the stipe, which is conspicuously spindle-shaped. *C. integerrimus*, another similar *Myxacium*, has smaller spores.

185 CORTINARIUS METRODII

Etymology After the French mycologist Metrod.
Description Cap 4–7 cm, yolk-yellow or orange-yellow, con-
ical-campulate then convex and also slightly depressed at disc,
smooth, viscid in wet weather. Gills light violet then rust-colored, thin, crowded, straight, adnate or slightly decurrent.
Stipe 7–9 × 1–1.5 cm, whitish, long, club-shaped or with a
distinctive bulb at base, which may also narrow into a point,
often recurved, covered by a whitish glutinous sheath, solid
then hollow. Veil whitish, cobwebby, fugacious. Flesh whitish
turning slightly yellow, especially at base. No particular odor or
flavor. Spores rusty brown, elliptical, warty, 10.5–11.5 ×
5.5–6.5 microns.
Edibility Uncertain.
Habitat In coniferous woods, very rare.
Season Late summer and autumn.
Note Other *Myxacium* fungi which may have the cap tinged
with orange are *C. vibratilis* (small, stipe slender, gills cream-
colored with very bitter flesh), and *C. collinitus* (stipe almost
cylindrical and pale blue in color, with pale gills).

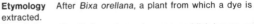

186 CORTINARIUS ORELLANUS

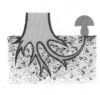

Etymology After *Bixa orellana*, a plant from which a dye is
extracted.
Description Cap 3–8 cm, tawny brown or reddish-brown, or
reddish-orange, convex-campanulate then flattened with a low,
obtuse umbo, finely feltlike, dry. Gills tawny-saffron then rusty-
brown, distant, thick, broad, adnexed or decurrent with a small
tooth. Stipe 4–9 × 1–2 cm, yellow then saffron-yellow, cylindrical or narrowing at base, fibrillose with reddish or cap-colored fibrils, soft. Veil pale yellow, cobwebby, fugacious. Flesh
light yellow then tawnyish, thin at cap margin. Slight odor of
radish, sweet flavor. Spores rusty brown, elliptical, warty,
8.5–12 × 5.5–6.5 microns.
Edibility Deadly poisonous.
Habitat Broadleaf woods, sometimes pine. Not common, but
has a wide distribution in Europe; reported, but not known for
certain, to occur in North America.
Season Late summer and autumn.
Note The number of poisonous or suspected poisonous *Cortinarius* fungi is increasing. *C. orellanus* belongs to the subgenus *Leprocybe*, as does lethal *C. speciosissimus* (**178**).

187 HEBELOMA CRUSTULINIFORME

Etymology From Latin, "biscuitlike," from its color.
Description Cap 5–15 cm, light ochre, yellowish, or brick-colored, disc a deeper color, brownish ochre, conical-convex, convex-flat, typically with large obtuse umbo, or slightly humped; when mature sometimes raised at margin, smooth, initially slightly viscid in damp weather. Gills whitish then ochreous-brown, eventually brown, short, sinuate, crowded, with edge denticulate, irregular and lighter. The gills produce small watery droplets in young specimens and in damp weather; when dry they look spotted. Stipe 4–7 × 1–2.5 cm, white, finely scaly toward top (sometimes to base also), solid, cylindrical or enlarged at base, almost bulbous. Flesh whitish. Rootlike odor and slightly bitter flavor. Spores brownish, elliptical, speckled, 10–12 × 5–6 microns.
Edibility Sometimes mildly poisonous.
Habitat In groups, beneath broadleaf species, especially poplars and birch, also under conifers.
Season Summer and autumn.
Note In North America there are over 100 species of *Hebeloma;* very few are well known and **none should be eaten.** The genus is recognized by the nonscaly, slimy cap, becoming shiny when dry, ochre-brown gills, often radishy odor, and its terrestrial habitat.

188 HEBELOMA SACCHARIOLENS

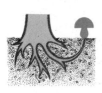

Etymology From Latin, "smelling of sugar."
Description Cap 2–6 cm, whitish, brown at disc, entire surface darkening with age, campanulate then convex, smooth, quite viscid in damp weather. Gills whitish then brown-ochre, eventually rust-colored, adnate, long, edge whitish, split. Stipe 4–5 × 1 cm, white, then tinged with ochre or brownish starting at base, fibrillose, white and pruinous at the top, full, narrowing toward top. Flesh yellowish, sometimes brown-ochre. Distinctive odor of orange blossom or caramel, flavor sweetish. Spores brownish, elliptical, finely speckled, 12–17 × 7–9 microns.
Edibility Not known.
Habitat In woodland, sometimes in grassland.
Season Summer and autumn.
Note The carpophores of *H. anthracophilum* form among charcoal remains; the flesh is elastic and bitter, stipe floccose, whitish and darkening from base upward. *H. sarcophyllum* grows in warm temperate regions; it is whitish, glabrous and smooth, with pale purple-pink gills and reddish spores, which are exceptional for this genus.

189 MYXOCYBE RADICOSA

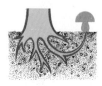

Synonym *Hebeloma radicosum.*
Etymology From Latin, "with a root."
Description Cap 5–15 cm or more, whitish ochre-brownish, never uniformly colored, also with reddish markings, in wet weather covered with an abundant shiny viscous layer tending to dry off revealing adpressed fibrillose scales, fleshy, convex then fairly expanded, edge faintly sulcate or recurved. Gills whitish then rust-brown, rounded at base, semifree, crowded. Stipe 7.5–25 × 1–2.5 cm, whitish with inferior membranous ring, white then brownish, above which stipe is white and mealy, below which covered with floccose scales that become reddish brown, sturdy, enlarged at base which extends into a long rooting appendage, completely solid. Flesh whitish, solid. Pleasant odor of bitter almonds or cherry laurel, flavor sweetish. Spores ochreous, elliptical, slightly roughened, 9–10 × 5–6 microns.
Edibility Can be eaten, although of mediocre quality.
Habitat Solitary or gregarious, in humus-rich woods.
Season Summer and autumn.
Note A large and easily identifiable species in Europe, its presence and distribution in North America are uncertain. A related genus, *Phaeocollybia*, has deeply rooting stipes under conifers along the Pacific Northwest Coast: they have rust-brown spores and lack a veil.

190 PHOLIOTA ADIPOSA

Etymology From Latin, "fat."
Description Cap 3–17 cm, yellow, covered with floccose-gelatinous surface scales, rust-colored on yellow background, concentrical and dropping off, cap thus eventually flat, sometimes humped, viscid in wet weather. Gills yellowish then rust-colored, adnate, broad. Stipe 6–16 × 1–2.5 cm, first whitish then yellow, covered beneath the ring by pointed rust-colored scales, gelatinous-viscous, enlarged toward base. Membranous ring, yellow then rust-colored. Flesh yellowish, light brown at base. Odorless, with a slightly bitter flavor. Spores rust-brown, elliptical, smooth, 5–7 × 3–4 microns.
Edibility Mediocre.
Habitat On trunks, often still living, of broadleaf species such as beech, maple, and birch, in mountainous areas.
Season Summer and autumn.
Note In *P. aurivella* the cap scales are entirely adpressed on rusty yellow background, broad and brownish. The stipe is not even slightly viscous, but fibrillose and floccose, whitish at the top tinged with yellowish-brown from the base upward.

191 PHOLIOTA SQUARROSA

Etymology From Latin, "covered with scales."
Description Cap 3–10 cm, rust-yellow or ochre, dry, convex-campanulate, flattened, barely raised at the center or humped, covered with evident warts, which are crowded, tall, persistent, darker. Gills pale green then rust-colored, adnate, crowded. Stipe 7–20 × 1–2.5 cm, cap-colored, narrowing toward the base, though sometimes enlarged at base, covered with scales, raised, crowded, dark brown below ring. Ring descending, fibrillose-frayed, dark brown. Flesh pale yellow. Odor of rotting wood, sometimes odorless, flavor not unpleasant. Spores rusty brown, elliptical, smooth, 6–8 × 3.5–4 microns.
Edibility Can be eaten, but see **caution.**

Habitat In dense tufts on tree stumps, trunks, or roots of broadleaf species, less often on conifers.
Season All year.
Note *P. squarrosoides* has erect to recurved cap and stalk scales, and is somewhat slimy in wet weather. *P. squarroso-adiposa*, with viscid yellow cap, has adpressed cap scales, but raised scales on the stipe.
Caution Some people experience mild to severe digestive upset.

192 PHOLIOTA ALNICOLA

Etymology From Latin, "living on alder."
Description Cap 3–11 cm, bright yellow and almost transparent when wet, also mucilaginous, ochreous when dry tending to become reddish or tinged with green, fleshy, convex then flat, margin initially fibrillose. Gills ochreous brown, pale then rust-colored, slightly adnate. Stipe 4–9 × 0.4–1.2 cm, lemon-yellow then invariably darkening from base, fibrillose or slightly floccose, top pale yellow eventually darkening, tapering toward bottom if tufted, otherwise slightly enlarged, usually curved or supple. Veil abundant, pale, fibrillose, remaining mostly adherent to cap edge. Flesh cap-colored, soft, rather fibrous in stipe. Odor pleasant, with a hint of spun sugar, flavor bitter. Spores rust-red, elliptical, 8–10 × 4–5.5 microns.

Edibility Inedible because of the bitter flavor.
Habitat Isolated but more often clustered on dead alder stumps and trunks, also on other broadleaf species in damp places.
Season Summer and autumn.
Note *P. flavida* is very similar, possibly identical, and grows on conifers.

193 PHOLIOTA DESTRUENS

Etymology From Latin, "destroying."
Description Cap 6–20 cm, yellowish white tending to brown-ochre at disc, covered with white scales, wooly, fugacious, sometimes slightly viscid, convex then flat, slightly umbonate or rarely humped, margin fibrillose, initially involute. Gills white becoming gradually brownish, rounded at stipe or fairly adnate, crowded. Stipe 5–17 × 2–3 cm, cap-colored, darker at enlarged, rooting stipe, covered by same type of scales, even if raised, below a white ring, floccose, fugacious. Flesh white, cinnamon-brown at stipe. Odor of malt, flavor somewhat bitter, but sometimes sweetish. Spores brownish, elliptical, smooth, 7–10 × 4–6 microns.
Edibility Edible, but of poor quality.
Habitat On dead broadleaf trunks, especially poplars.
Season Summer and autumn.
Note This is commonly seen where poplar has been cut. Its mycelium readily penetrates and breaks down the soft wood, and mushrooms develop on stumps and out of the cut surfaces of logs.

194 PHOLIOTA CARBONARIA

Etymology From Latin, "of coal," because of its habitat.
Description Cap 3–9 cm, ochreous red, orange-brown, viscid in damp weather, fairly shiny when dry, glabrous or finely squamulose toward edge, recurved, convex then flat and slightly depressed at disc. Gills clay-colored, then tinged with pale straw-yellow eventually turning grayish brown, adnate, crowded. Stipe 2.5–11 × 0.2–1.4 cm, whitish or pale lemon-yellow at top, reddish lower-down, blackish at base, elastic and tough, cylindrical or tapering toward base, fibrillose and squamose, with a mycelium that incorporates the substratum in a spheroidal mass at the base. Veil fibrillose and fugacious. Flesh yellowish, whitish toward stipe, firm. Slightly earthy odor, flavor sweetish. Spores dark rusty-brown, elliptical, smooth, 6–8 × 3.5–4 microns.
Edibility Edible, but of poor quality.
Habitat Isolated or in small tufts on the remains of charcoal kilns or fires, or in burnt ground.
Season Early spring to the first winter frosts.
Note This fungus is easily identified by its habitat, which also hosts other fungi such as *Hebeloma anthracophilum, Geopetalum carbonarium* and some cup-shaped Ascomycetes belonging to the genera *Anthracobia, Geopyxis, Peziza,* and *Ascobolus.*

195 AGROCYBE AEGERITA

Synonyms *Pholiota aegerita; Pholiota cylindracea.*
Etymology From Greek, "pertaining to the black poplar."
Description Cap 3–12 cm, pale ochre-brown tending to fade to whitish from margin, convex then rugose and flat, slightly viscid and opalescent if wet, silky when dry, often cracked and areolated at disc. Gills whitish then grayish-brown, adnate or adnate-decurrent, fairly crowded. Stipe 8–15 × 1.5–3 cm, whitish tinged with pale ochre-brown, narrowing toward base, fistular, fragile, and fibrillose. Ring white, membranous, superior, pendant. Flesh whitish, brown beneath cuticle and at base. Pleasant cheeselike odor and flavor. Spores brown, broadly elliptical, smooth, 8–10 × 5–7 microns.
Edibility Very good.
Habitat In tufts or gregarious on dead wood, stumps and trunks of broadleaf species (prefers poplars), also on living wood.
Season Most frequent from spring to late autumn.
Note Well-known since antiquity, this was probably the first fungus to be cultivated artificially. Easily grown using small poplar trunks, straw or other vegetable remains, in the right conditions it will form its first fruit-bodies in 2–3 months. In order to bear fruit the mycelium needs sunlight.

---

196 KUEHNEROMYCES MUTABILIS

Synonym *Pholiota mutabils.*
Description Cap 3–6 cm, reddish to dark brown, very hygrophanous when wet, fading to ochreous on drying, smooth, convex then flat or faintly umbonate, sometimes depressed or irregularly humped when old. Gills pale yellow then cinnamon, adnate-decurrent, crowded, quite broad. Stipe 4–8 × 0.5–1 cm, rust-brown, blackish toward base, covered up to ring with raised scales, smooth or very pale at top, rigid, cylindrical or narrowing toward base, curved. Ring membranous, brownish, persistent. Flesh whitish. Strong pleasant odor, flavor sweetish. Spores brownish ochre, ovoid, smooth, 6–7.5 × 3–5 microns.
Edibility Caps very good, but see **note.**
Habitat In tufts, sometimes quite extensive, on broadleaf and conifer stumps and logs.
Season From spring to first winter frosts.
Note There is a **deadly look-alike,** also quite tasty. *Galerina autumnalis* and related species grow on decaying logs, in groups but not in dense tufts; stipes lack the raised or recurved scales, and spores are ornamented. Poisoning is the same as that caused by the destroying angel.
Caution See note.

197 ROZITES CAPERATA

Synonym *Pholiota caperata.*
Etymology From Latin, "wrinkled," from cap appearance.
Description Cap 5–12 cm, yellowish, fleshy, soft, first campanulate then convex, eventually flat, shiny, cuticle with scattered silvery pruinescence, detaching along edge, persistent at center, thin irregular grooves, especially at the margin, making it slightly wrinkled or puckered. Gills yellowish, crowded, of average length, adnate. Stipe 6–15 × 1–2.5 cm, whitish cream, sturdy, fibrous, solid, cylindrical or narrowing from bottom to top, with soft, membranous, yellowish white ring. Flesh creamy white, soft and fragile. Odor faint, flavor sweet. Spores rusty brown, elliptical, warty, 11–14 × 7–9 microns.
Edibility Good; turns water yellow when boiled.
Habitat In groups in woods, sometimes broadleaf, but particularly pine and fir, in moss and bilberries, among stones and on dry, gravelly, acid ground.
Season Late spring to late autumn.
Note Although superficially resembling an *Agrocybe* or *Pholiota* in appearance, this mushroom, the only North American species in this genus, is classed separately because of its slightly roughened-warty spores.

198 GYMNOPILUS SPECTABILIS

Synonyms *G. junonius; Pholiota spectabilis.*
Etymology From Latin, "notable" or "remarkable."
Description Cap 5–15 cm or more, semiglobose, fairly convex, expanded, slightly undulate, compact, fleshy, with fibrils or innate small fibrillose scales, adpressed, orange-yellow against a golden yellow background, eventually leather-yellow. Gills yellow then rust-colored, adnate, very crowded. Stipe 6–13 × 2–3 cm or more, sulfur-yellow or cap-colored, sturdy, fairly enlarged at middle, normally with rooting base, fusiform, sheathed by veil, sometimes squamulose, fibrillose, sometimes smooth, shiny, mealy at top, with full membranous ring, open then adherent, yellowish then rust-colored, inferior, persistent. Flesh sulfur-yellow, reddish when touched, thick, compact. Pleasant odor, bitter but aromatic flavor. Spores rust-colored, almond-shaped, warty, 7.5–9 × 5.5–7 microns.
Edibility Eaten in Europe, but quite bitter; reported to cause transient alteration in visual perception.
Habitat Tufted, on stumps or at the base of unhealthy coniferous and broadleaf trees.
Season Summer and autumn.
Note This widely distributed species is more common on some trees than others, depending on the environment, although it can adapt to any type of wood.

199 PHAEOLEPIOTA AUREA

Synonym *Pholiota aurea.*
Etymology From Latin, "gold-colored."
Description Cap 4–15 cm, golden ochre-yellow, first powdery because of small and quite crowded scales, then velvety and darker, hemispherical then convex, eventually expanded with inconspicuous umbo. Gills rounded toward stipe, ochreous then rust-colored, crowded. Stipe 6–28 × 1–3.5 cm, sturdy, almost cylindrical, slightly enlarged at base, full, with large ring, ascendant, permanent, formed by extension of veil which sheaths much of stipe, and which is easily detachable down to base, almost cap-colored, similarly powdery and then squamulose-velvety, whitish above ring, pale ochre and pruinous. Flesh soft, whitish and pale yellow when exposed to air, reddish toward base. Strong, aromatic odor, distinctive flavor. Spores golden ochre, elliptical, roughened, 9–15 × 4–6 microns.
Edibility Good, but it upsets some stomachs.
Habitat In groups near alder in the Pacific Northwest.
Season Summer and autumn.
Note This is a large, uncommon mushroom, resembling a cross between a *Pholiota* and a *Cystoderma*.

200 CONOCYBE TENERA

Etymology From Latin, "tender" or "delicate."
Description Cap 1–2 cm, pale rust-colored fading when dry, hygrophanous, thin, conical-campanulate, smooth. Gills cinnamon-colored, adnexed or free, straight and quite crowded. Stipe 7.5–10 × 1.5–2 cm, cap-colored, fragile, cylindrical, straight and rigid, silky, striate lengthwise, powdery. Flesh membranous in cap, slightly fibrous in stipe, but almost nonexistent. Spores pale rust-colored, elliptical, smooth, 8–15 × 5.5–8 microns.
Edibility Of no interest because of the size, and possibly poisonous, like some species in the genus.
Habitat In soil in woods, grassland, and gardens.
Season From spring to first winter frosts.
Note It belongs to a group of small fungi of no gastronomic interest; most of them require microscopic examination to establish the precise species. One common spring lawn species, *C. lactea*, has a white cap and stalk and cinnamon gills; it comes up overnight after rains and usually disappears by noon.

201 PHOLIOTINA TOGULARIS

Synonyms *Conocybe togularis; Philiota togularis.*
Etymology From Latin, "with a small cloak."
Description Cap 0.5–2 cm, pale ochre, darker at disc, campanulate then expanded, slightly convex, sometimes striate. Gills yellow, light rust-colored when mature, crowded, narrowing at both ends. Stipe 1–4 cm × 0.5–1.5 mm, yellowish tending to darken toward base which is enlarged, fibrillose lengthwise, pruinose at top. Ring whitish, membranous, central, broad. Flesh very thin, pale yellow. Spores rust-ochre, smooth, varying in size because they are produced by two- and four-spored basidia, by the former 10.5–12.6 × 5.4–6.8 microns, in the latter 8–10 × 4.5–5.5 microns.
Edibility Possibly deadly poisonous.
Habitat In grassland, gardens, and pastures.
Season Spring to late autumn.
Note Many small gilled fungi are called "little brown mushrooms"; none is readily recognized in the field, and one (*Conocybe filaris*) has recently been found to contain amanitins, toxic compounds causing the lethal poisoning by the destroying angel group of *Amanitas.*
Caution Avoid all "lbm's": little is known about the edibility of most of them, and several unrelated species are now known to cause serious poisoning.

202 AGARICUS AUGUSTUS

Common name The prince.
Synonym *Psalliota augusta.*
Etymology From Latin, "majestic."
Description Cap 10–20 cm, initially subglobose, flat at top, then convex and eventually flattened, cuticle dry and detachable, broken up into small fibrillose scales, adpressed, reddish brown on a yellow-cream background. Gills crowded, free, white then gray, pink and eventually chocolate-brown. Stipe 10–20 × 2–3 cm, cylindrical, enlarged at base, solid then slightly hollow, white with floccosity turning yellow, pinkish above ring, which is large, membranous, and double with brownish enlargement in lower part. Flesh white, turning yellow then brown when exposed to air, reddish at end of stipe. Odor of almonds, flavor sweet. Spores brown, ovoid, smooth, 7.5–9 × 5–5.5 microns.
Edibility Excellent.
Habitat In grassy areas near conifers.
Season Late summer and autumn.
Note The gills, which remain white for a long period, may cause it to be confused with a *Lepiota.* It is most common from the Rocky Mountains to the Pacific Coast, and is one of several large, fleshy species of *Agaricus* which both bruise yellow and have a pleasant, almondlike or aniseed odor.

203 AGARICUS XANTHODERMUS

Etymology From Greek, "yellow-skinned."
Description Cap 6–12 cm, globose-cylindrical then truncated-conical, finally convex but flattened at top and expanded; cuticle dry, white, at times tinged grayish or brownish, or with small adpressed scales; normally the surface has chrome-yellow markings. Gills free, crowded, whitish then pinkish, eventually blackish brown. Stipe 8–15 × 0.8–1.2 cm, cylindrical, often curved toward base, with conspicuous basal bulb, fistular, smooth, silky, white but chrome-yellow when touched, especially lower down. Ring membranous, inferior, slightly floccose and denticulate in lower part, white with margin turning yellow then darkening. Flesh white, turning slightly yellow beneath cuticle, decidedly chrome-yellow at base. Strong odor of carbolic acid, iodine, or ink, flavor sweetish. Spores blackish brown, ovoid, smooth, 5–6.5 × 3.5–5 microns.

Edibility Slightly poisonous, usually causing gastrointestinal distress lasting a few hours to a day.
Habitat In groups, in open woods or grassland, meadows, and gardens.
Season Summer and autumn.
Note All *Agaricus* species with foul-smelling gills are known to cause stomach upsets.

204 AGARICUS ARVENSIS

Synonym *Psalliota arvensis.*
Common name Horse mushroom.
Etymology From Latin, "of fields," from its habitat.
Description Cap 7–15 cm, subglobose becoming flattened, silky, white, turning yellow when touched, frayed with velar remains at margin. Gills crowded, grayish turning pinkish and eventually blackish, with white edge, free. Stipe 8–13 × 1.5–3 cm, club-shaped, fistular when mature, white, turning yellow, ring descendant, double with toothed crown lower down. Flesh white, with age or when drying out becoming ochre-yellow starting from the base. Odor of aniseed, flavor sweet. Spores dark brown, oval, smooth, 7–8 × 4–5 microns.

Edibility Excellent.
Habitat In groups of fairy rings in fields, pastures, grassy areas.
Season Summer and autumn.
Note Some similar species which grow in woodland are often confused with *A. arvensis* because of their appearance and the smell of aniseed.

205 AGARICUS SILVICOLA

Synonym *Psalliota silvicola.*
Common name Wood mushroom.
Etymology From Latin, "growing in woods."
Description Cap 5–12 cm, first globose then campanulate becoming expanded, cuticle dry, shiny, whitish, yellowing with age or when touched. Gills crowded, free, initially dirty white then pinkish, sepia-brown when mature. Stipe 6–11 × 1.5–2.5 cm, slender, faintly hollow, fragile, basal bulb white turning to yellow, pinkish above ring, ring double with darkening fringe. Flesh white, pinkish in stipe. Faint smell of aniseed, flavor pleasant. Spores cocoa-colored, oval, smooth, 5–6 × 3–4 microns.
Edibility Good, but see **caution.**
Habitat In sunny coniferous woodland, sometimes also broadleaf.
Season Summer and autumn.
Note *A. abruptibulbus* is a very similar species, but the bulb is clearly marginate and slightly curved at the base, and the spores are larger.
Caution When collecting woodland species of *Agaricus,* be especially careful with buttons or immature specimens: they can be confused with the destroying angel group of *Amanita.*

206 AGARICUS BISPORUS

Common names Cultivated mushroom; champignon.
Synonym *Agaricus brunnescens.*
Etymology From Latin, "with two spores."
Description Cap 5–10 cm, white then rose, brownish when mature, fleshy, globose or hemispherical then convex and also completely expanded, flat, young specimens with white, soft, denticulate, fugacious fringe at edge. Gills rose-white in young specimens, reddish brownish in mature fungi, crowded, not adherent. Stipe 3–5 × 1–1.5 cm, white, sometimes rose in young specimens above ring, squat, heterogeneous, pithy then somewhat hollow; ring white, thick, soft, membranous, descending. Flesh white, when exposed to air young specimens are tinged rose, older specimens tinged brownish, thick, soft. Pleasant odor and flavor. Spores cocoa- or violet-brown, elliptical, smooth, 6–9 × 4–6 microns.
Edibility Excellent.
Habitat In grassland, gardens, orchards, meadows, beside roads, in open places, on horse droppings.
Season Late spring to autumn.
Note Cultivated since the 18th century and given rise to a flourishing agricultural industry in all countries with a temperate climate. This is the only fresh mushroom sold throughout North America; other kinds can sometimes be found locally in gourmet shops.

207 AGARICUS LANGEI

Etymology After the Danish mycologist Lange.
Description Cap 6–12 cm, semiglobose then expanded, covered with brownish fibrillose scales, thinning out toward edge, cuticle joined and darker at center, dry. Gills free, pinkish, finally blackish brown, crowded, short, margin sterile and whitish. Stipe 7–12 × 1.5–2.5 cm, almost cylindrical or faintly enlarged at base, smooth above ring, white and floccose below, tinged with red and then darkening if touched, hollow. Ring descending, large, membranous, with small brown scales below. Flesh at once crimson when cut. Slightly acid but pleasant odor and flavor. Spores blackish brown, ovoid, smooth, 7–9 × 3.5–5 microns.

Edibility Good.
Habitat Gregarious, in coniferous woods (prefers Norway spruce), rarely, beneath broadleaf species.
Season Summer and autumn.
Note Although *A. langei* is not found in North America, several red-staining species of *Agaricus* are. Difficult to identify to species, some are edible, but others are reported to cause gastrointestinal distress. *A. haemorrhoidarius* has smaller spores, brownish scales on stipe, and grows in broadleaf woods. *A. silvaticus*, under conifers and in mixed woods, bruises red more slowly.

208 STROPHARIA AERUGINOSA

Etymology From Latin, ''copper-green.''
Description Cap 3–7 cm, bluish green, tending to turn yellow or pale as viscous, detachable cuticle dries up; while water-filled, bluish, and flesh beneath, yellowish; rather fleshy, campanulate-convex, then flat, slightly umbonate or obtuse, often squamulose, whitish and fugacious. Gills white then brownish sometimes finally purplish brown, adnate, not very crowded. Stipe 4–10 × 0.4–1.2 cm, cap-colored though paling more quickly, white, caducous. Ring white uppermost, stipe-colored lower down, floccose, fugacious. Flesh bluish, fading to whitish. No odor or flavor. Spores violet-brown, elliptical, smooth, 7–9 × 4–5 microns.

Edibility Quite good without the viscous cuticle.
Habitat On litter in broadleaf woods, sometimes under conifers, and in open grassy areas.
Season Late spring, summer, and autumn.
Note The pale green coloration also marks the small *S. albocyanea*, green then whitish with colorless mucus, stipe not viscid, whitish or pale green, with incomplete white ring. *S. cyanea* is much more similar, ring floccose and evanescent, gills brownish with lighter edges. Hallucinogenic *Psilocybe* species turn blue on bruising.

209 PSILOCYBE SQUAMOSA

Synonyms *Stropharia squamosa; Naematoloma squamosum.*

Etymology From Latin, "scaly."

Description Cap 3–5 cm, hemispherical or campanulate, eventually almost flat, hygrophanous, olive-gray, orange-yellow at disc, turning straw-yellow when drying, starting at edge; detachable viscous layer, with vaguely concentric pale yellow scales, fugacious, initially making margin appendiculate. Gills violet then dark gray, white-edged, quite crowded, adnate. Stipe slender, 3–16 × 0.3–0.8 cm, rigid, straight to curved at slightly enlarged base, covered up to frayed membranous ring by a fibrillose veil, broken up into small raised scales, yellowish then reddish at base, white and finely pruinous-striate at top, fistular. Flesh pale yellow in cap, fibrous and reddish-yellow toward stipe. No odor or flavor. Spores violet-brown, elliptical, smooth, 12.5–14.5 × 7–8 microns.

Edibility Edibility uncertain.

Habitat In broadleaf litter, basal mycelium closely associated with vegetable remains, dead wood, or leaves.

Season Summer and autumn.

Note *P. thrausta* is very similar but has a bright reddish cap. Although both mushrooms are now considered to be species of *Psilocybe*, neither is hallucinogenic.

Caution Edibility is uncertain.

210 STROPHARIA RUGOSOANNULATA

Synonym *Stropharia ferrii.*

Etymology From Latin, "wrinkled and ringed."

Description Cap 5–20 cm, ochreous to brick-red with pale purple shades, first semiglobose, convex with involute margin, then flattened, very often split at margin, fleshy, fibrillose, dry. Gills violet-gray, adnate, crowded, thin. Stipe 7–14 × 0.9–1.5 cm, cylindrical, faintly narrowing toward top, solid, slightly tinged with yellow lower down, large membranous ring, sulcate higher up, where it soon becomes colored by spores. Flesh white, soft. No odor or particular flavor, sometimes slightly earthy. Spores violet-charcoal-gray, elongated-ovoid, smooth, 10–12 × 6–8 microns.

Edibility Delicious.

Habitat In wood chip mulch and humus-rich soils.

Season Spring to late autumn.

Note Easy to cultivate on ground with plenty of straw, this robust, choice edible is found in the mulch around garden shrubbery in abundance. It is easy to confuse with some species of *Agaricus* (but they have free gills) and some species of *Agrocybe* (but they have brown gills and give a brown spore print).

211 STROPHARIA SEMIGLOBATA

Etymology From Latin, "like a half sphere."
Description Cap 1–5 cm, light yellow, hemispherical, obtuse, smooth, viscid. Gills very broad, up to 1.3 cm long, flat, adnate, grayish with black marks. Stipe 5–15 × 0.2–0.4 cm, pale yellow, paler at top, straight, rigid, faintly enlarged at base, covered by a glutinous transparent veil beneath ring which is thin, viscid, and incomplete. Flesh pale yellow, thin. Odor mealy, flavor sweetish. Spores violet-brown, elliptical, smooth, 16–21 × 8–10 microns.
Edibility Of no interest because of size.
Habitat Directly on manure, mainly of horses or cattle, and among tall grass in meadows.
Season Spring to late autumn.
Note This is one of the many dark-spored, gilled fungi that grow on dung, including species of *Coprinus* (**227**), *Panaeolus* (**217, 218**), *Anellaria* (**219**), and *Psilocybe*. Their identification to species is often difficult without microscopic examination. One common coprophilous *Psilocybe* in southern states is the hallucinogenic *P. cubensis*, which is larger, has a ring, and bruises blue.

212 STROPHARIA CORONILLA

Etymology From Greek, "with a small garland."
Description Cap 2–5 cm, ochreous yellow, lemon-yellow, fleshy, hemispherical then convex, eventually flat, smooth with white, floccose margin. Gills pale yellow then violet, white-edged, adnate, crowded. Stipe 3–4 × 0.4–1 cm, white turning yellowish with age or when touched, cylindrical, ring white, persistent, striate, in central part of stipe. Flesh white and soft. No particular odor or flavor. Spores violet-brown, elliptical, smooth, 6.5–10 × 4–6 microns.
Edibility Possibly mildly toxic.
Habitat Common in lawns and grassy areas.
Season Spring to late autumn.
Note Resembles a small field mushroom (*Agaricus*), but the gills are not free. *S. melanosperma* is very similar, but rarer, with cap whitish, pale yellow at disc, stipe white and striate at top, ring well developed and spores blackish-violet. Another species, *S. inuncta*, is first yellowish then purplish red on cap, which is covered with a viscous, detachable layer. The supple white stipe, with thin ring, is pruinous at top, dry and fibrillose beneath ring. It grows in fields. Similar is *S. albonitens*, which is completely white in the stipe and has a viscous cap.
Caution Possibly toxic.

213 NAEMATOLOMA CAPNOIDES

Synonym *Hypholoma capnoides.*

Etymology From Greek, "smoky" or "smoke-colored," referring to the gills.

Description Cap 2.5–8 cm, ochreous yellow, convex then flat, dry, smooth, sometimes rugose, margin appendiculate with whitish velar remains. Gills whitish then ash-gray, eventually violet-gray, adnexed, easily detachable from flesh, fairly crowded. Stipe 5–7 × 0.4–0.8 cm, pale ochre, then rust-brown and darker at base, top whitish, cylindrical, normally curved and supple, unevenly striate, veil white then violet-brown. Flesh whitish, often rust-colored toward base of stipe. No particular odor or flavor. Spores violet-brown, elliptical, smooth, 7–8 × 3–4 microns.

Edibility Fair to good.

Habitat Normally tufted on coniferous wood.

Season From spring to the first winter frosts.

Note This is the tastiest of the group. Unlike the other more common species, it grows exclusively in coniferous wood. Another species, *N. sublateritium,* has brick-red caps and grows on logs and stumps of broadleaf wood; it is a good edible when young.

---

214 NAEMATOLOMA FASCICULARE

Synonym *Hypholoma fasciculare.*

Common name Sulfur Tuft.

Etymology From Latin, "in small bundles."

Description Cap 2–5 cm, bright yellow, often darker at center, convex then flat, rarely with a slight raised area at disc, smooth, dry margin sometimes appendiculate with velar remains. Gills sulfur-yellow then pale green, grayish brown when mature, adnexed, crowded, almost deliquescent. Stipe 5–22 × 0.4–1 cm, yellow, tapering or enlarged at base, curved or supple, fibrillose toward base, veil first pale yellow, cobwebby, then covered with spores. Flesh thin and yellow. Faint distinctive odor, flavor bitter. Spores violet-brown, elliptical, smooth, 5–7 × 3.5–4 microns.

Edibility Inedible, possibly poisonous.

Habitat On stumps or buried roots of both conifers and broadleaf species, sometimes apparently parasitic, normally tufted.

Season From spring to first winter frosts.

Note *N. fasciculare* is easily recognized by its clustered growth pattern, the greenish tint of its gills, and its bitter taste. In Europe and Asia there are reports of serious, even fatal poisonings from eating this species.

215 NAEMATOLOMA DISPERSUM

Synonym *Hypholoma dispersum.*
Etymology From Latin, "scattered."
Description Cap 2–4 cm, honey-brown, campanulate then convex, finally flat, smooth, with long white velar fragments on the surface. Gills pale straw-colored, eventually slightly pale greenish gray, adnexed, ventricose, crowded, white-edged. Stipe 5–7 × 0.4–0.6 cm, cylindrical, erect, tough, sometimes rust-brown, darkening at base, pale at top, striped with white, silky zonations of velar remains. Flesh yellowish, darker beneath cuticle and at base. No discernible odor, very bitter flavor. Spores violet-brown, oval, smooth, 7–9 × 4–5 microns.

Edibility Inedible.
Habitat Gregarious on or around tree stumps, rarely on sawdust, preferably on conifers, especially common along logging roads in the Pacific Northwest.
Season Spring to early winter.
Note Unlike other *Naematolomas*, this mushroom is not tufted; rather, it grows in large, unattached groupings.

216 PANAEOLINA FOENISECII

Synonyms *Panaeolus foenisecii; Psathyrella foenisecii.*
Etymology From Latin, "dry hay."
Description Cap 1.5–3 cm, first almost hemispherical, convex or campanulate-convex, very rarely expanded and slightly umbonate, glabrous, smooth, hygrophanous, soot-gray discoloring when drying out, starting from disc becoming light brownish ochre, slightly cracked in very dry weather. Gills grayish, blackish brown when mature, with grayish white edge, broad, quite distant, soon detached from stipe, ventricose toward margin. Stipe 3.5–10 × 0.2–0.3 cm, brownish gray then light ochreous, whitish, silky, and fibrillose at top, pubescent at base, cylindrical, fistular, cartilaginous but fragile. Flesh thin, fragile, pale brownish ochre in dry weather, grayish brown when wet. Mushroomy odor, slightly acid flavor. Spores dark brown, not blackish, almond- or lemon-shaped, warty, 12–16.5 × 7.5–9 microns.
Edibility Slightly hallucinogenic in eastern North America when ingested in large quantities.
Habitat Gregarious in grassy areas.
Season Late spring through early autumn.
Note This is one of several lawn fungi that come up after spring rains.

217 PANAEOLUS ATER

Etymology From Latin, "black."
Description Cap 1–1.5 cm, subglobose, convex, obtuse, dark olive-brown, sepia-black, fading in dry weather from margin, turning ochreous gray-brown sometimes tinged reddish, faintly striate at margin in damp weather. Gills adnate, blackish gray to black, ventricose, not very crowded, edge speckled with grayish. Stipe 3–7 × 0.1–0.3 cm, cylindrical, fistular, barely enlarged at top, finely striate, base pruinose, fistular, cap-colored but lighter. Flesh thin, dark brown in moist carpophore, brownish ochre when dry. No particular odor or flavor. Spores blackish, lemon-shaped, smooth, 10–12.5 × 6–7 microns.
Edibility Reportedly poisonous.
Habitat Isolated or in groups, on the droppings of herbivorous animals, or in fertilized grassland.
Season All year, except during frosts.
Note Carpophores are easily recognizable while dark, but they resemble *P. fimicola* when turning lighter (which can be distinguished by its reddish brown stipe, particularly near base). Some species of *Panaeolus* have been found to contain compounds similar to those found in some *Psilocybe* mushrooms.

218 PANAEOLUS CAMPANULATUS

Etymology From Latin, "bell-shaped."
Description Cap 2–4.5 cm, with two different forms: hemispherical-convex with no mamelon, white or light hazel; or conical-campanulate, campanulate-obtuse with mamelon, soot-brown. The cap form depends on color, age, and hygrophany of specimen. Cuticle smooth or scaly, cracked when dry in first form. Margin whitish, sometimes festooned and appendiculate with denticular velar remains, cuticle easily detachable in very wet specimens. Stipe 3–7.5 × 1.5–2.5 cm or more, slender, fistular, cylindrical, slightly enlarged at base, striate more conspicuously at top, pruinous, reddish gray, flesh-colored lower down, cream, ochreous at top. In first form stipe more white and pruinous, and shorter. Gills grayish speckled with black, white-edged, sometimes with droplets of exudate, fairly distant, adnate. Flesh faintly thicker in disc, but still thin, grayish or faintly reddish. No particular odor or flavor, but smells like burnt sugar when cut. Spores blackish, ovoid or lemon-shaped, smooth, 15–18.5 × 10–12.5 microns.
Edibility Difficult to distinguish and should be avoided.
Habitat Isolated or in small groups, on or near droppings, normally in fields and grassland.
Season Late spring to early winter.
Note This is most recognizable when the cap margin is adorned with white velar remains resembling a row of white teeth, or it can be identified by its mottled gills. Most *Panaeolus* species grow on dung and are difficult to distinguish.

219 ANELLARIA SEMIOVATA

Synonyms   *Panaeolus separatus; Panaeolus semiovatus.*
Etymology   From Latin, "like a half-egg."
Description   Cap 2–6 cm, whitish or clay-ochre, ovate-campanulate, never expanded, apex obtuse, viscid, shiny when dry, rugose and then also cracked when older, margin often appendiculate. Gills whitish, then ash-black, adnate but virtually separate, ascending, up to 0.8 cm long, often whitish-edged. Stipe 5–20 × 0.4–0.8 cm, whitish, rigid and straight, faintly enlarged at base, finely striate, smooth, ring white, membranous, narrow, persistent, sometimes striate, some distance from cap. Flesh whitish, pale yellow beneath cuticle or toward base of stipe. No particular odor or flavor. Spores blackish, elongated, smooth, 16–22 × 9–15 microns.
Edibility   Edibility uncertain.
Habitat   On droppings, especially of horses, in grassland and woodland.
Season   Spring to early winter.
Note   A large coprophilous species, easily identifiable because of its ring. *A. phalaenarum* is another large and very similar species, cap whitish tinged with reddish, also on the stipe, with no ring, and a smell of burnt sugar.

---

220 LACRIMARIA VELUTINA

Synonyms   *Hypholoma velutinum; Psathyrella velutina.*
Etymology   From Latin, meaning "velvety."
Description   Cap 5–15 cm, dirty ochre, campanulate then expanded with slight central relief, hygrophanous, for a long period covered with tufts of adpressed fibrils, margin appendiculate with white velar remains. Gills brown then blackish brown dotted with black, adnexed, easily detachable, not very crowded, white-edged, floccose, producing watery droplets. Stipe 5–13 × 0.4–1.5 cm, dark ochre, cylindrical, fistular, fragile, covered with silky fibrils, downy at top above velar remains. Veil white then black and wooly. Flesh whitish. Insignificant odor and flavor. Spores violet-brown, blackish, elliptical, warty, 8–12 × 5–7 microns.
Edibility   Good when young, but see **caution.**
Habitat   In woodland and grassland, sometimes tufted, common by roadsides and in excavated areas.
Season   Spring to late autumn.
Note   This is a common urban and suburban mushroom that comes up in large numbers in parks and grassy areas.
Caution   Because of its gross similarity to some poisonous species of *Inocybe*, a spore print should be made to be sure the spores are dark violet-blackish.

221 PSATHYRELLA SUBATRATA

Etymology From Latin, "almost black."
Description Cap 2.5–5 cm, campanulate then expanded, obtuse or almost umbonate, thin, reddish gray, soot-brown, olive-brown, turning light reddish when drying out, smooth, slightly striate at the margin. Gills blackish, fairly adnate depending on the state of expansion of the cap, linear or ventricose. Stipe 2.5–20 × 0.2–0.4 cm, quickly becoming whitish, straight, rigid, cylindrical, smooth. Flesh yellowish or dark brown in the cap, whitish in the stipe, thin. Odor undefined, flavor bitter. Spores soot-black, elliptical, smooth, 14–17 × 7–9 microns.

Edibility Of no interest because of size and bitterness.
Habitat Gregarious, in grassland and woodland, and under bushes.
Season Summer and autumn.
Note *Psathyrella* is a large genus comprising some two hundred species. Many grow in the same habitat and are too similar to differentiate without microscopic examination. These are mostly fragile mushrooms with brownish fading caps, black to brownish black spores, and delicate white stipes.

222 BOLBITIUS VITELLINUS

Etymology From Latin, "yolklike," from its color.
Description Cap 2–5 cm, oval becoming almost flat, cuticle viscous, yolk-yellow tending to fade from the edge, deeply striate-sulcate when mature. Gills ochreous or yellow-brown tinged pinkish, quite distant, semifree, edge frayed. Stipe 5–8 × 0.3–0.7 cm, almost cylindrical, fragile, hollow, pale yellow with white floccose elements at the base. Flesh thin, almost membranous, pale yellow. No particular odor or flavor. Spores light yellow-ochre, oval, almond-shaped, smooth, 12–14 × 7–8 microns.

Edibility Of no interest because slimy, small, and fleshless.
Habitat Gregarious on very decomposed dung, on compost, or in fertilized land.
Season Spring to autumn.
Note This is often found in large quantities in early spring and again in the fall, preferring cooler weather. It is readily identified by its viscid, striate yellow cap and its pinkish brown gills.

223 COPRINUS ATRAMENTARIUS

Etymology From Latin, "inky" or "pertaining to ink."
Description Cap 5–8 cm, soot-brown, lead-gray, adpressed scales, persistently marked, brown or brown-ochre, silky and shiny, ovate, obtuse then campanulate with lengthwise grooves and ribs, soft to touch, minutely pruinous when young, often squamulose at disc, darker, margin recurved when mature. Gills white then blackish brown, deliquescent, free, ventricose, up to 1.5 cm long, edge floccose, Stipe 7–20 × 0.8–1.8 cm, white, initially ventricose, fusiform, narrowing in lower part slightly, more at top, sulcate, lengthwise fibrillose, smooth at apex, with small brown scales at base, hollow, Flesh gray-brown, parting radially in cap, fibrous in stipe. No special odor or flavor. Spores black, elliptical, smooth, 7–11 × 5–6.5 microns.
Edibility Good, but avoid alcohol (see **note**).
Habitat In grassy areas, often tufted, usually associated with buried wood or roots.
Season Spring to early winter.
Note When eaten with or followed by alcohol (up to 24 hours later), some people experience a "poisoning" characterized by flushing in face and neck, tingling in fingers and toes, headache, and sometimes nausea.

224 COPRINUS COMATUS

Common names Shaggy Cap, Lawyer's Wig.
Etymology From Latin, "with dense hair," from its cap.
Description Cap 4–6 cm, white turning pink at margin then black, cylindrical when young, up to 20 cm tall, then campanulate, cuticle initially continuous then quickly breaking up into soft, broad, imbricate scales, often raised, white, then ochreous at margin, with top of cap nonsquamose, entire, and brown-ochre, striate lengthwise, margin often recurved and split when mature. Gills white then pink, finally black and deliquescent, free, straight, crowded, up to 1 cm long. Stipe 12–25 × 1–2 cm, white then dirty white or tinged pale lilac, narrowing toward top, with enlarged rooting base, hollow, with white, movable, thin, fugacious ring. Flesh white, fibrous, in stipe, soft in cap where it splits radially. No odor or flavor. Spores blackish, elliptical, smooth, 11–13 × 6–7 microns.
Edibility Very good in youth, while gills still white.
Habitat Isolated or gregarious, in woods, grassland, gardens, by roadsides, especially in fine dark ground, very common in urban and suburban areas.
Season Spring to early winter.

225 COPRINUS DISSEMINATUS

Synonym *Pseudocoprinus disseminatus.*
Etymology From Latin, "scattered."
Description Cap 1–2 cm, whitish or yellowish, then ash-colored, normally gray-brown, with the center yellowish, at first ovate, then parabolic, finally slightly expanded, furfuraceous, then glabrous and deeply striate-sulcate. Gills whitish then blackish, adnexed. Stipe 2.5–6 × 0.2 cm, white, fragile, sometimes supple, at first slightly furfuraceous owing to a silky white mycelium. Flesh very thin, whitish, faintly yellow at cap center. No odor or flavor. Spores blackish, elliptical, smooth, 9–10 × 5–6 microns.

Edibility Of no interest because of size and difficulty of cleaning.
Habitat In crowded groups, often tufted, on or around old broadleaf stumps or in leaves.
Season Spring to late autumn.
Note It occurs in groups of hundreds, often with successive generations being added in the span of a few weeks. Although it appears to be a typical *Coprinus* its gills do not become inky as it matures.

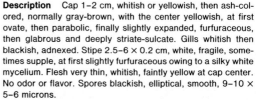

226 COPRINUS MICACEUS

Common name Inky cap.
Etymology From Latin, "micalike," because of the ornamentation of the cap when very young.
Description Cap 3–6 cm, rust-yellow, grayish yellow-brown, darker at center, finally grayish brown, ochre-rust when dry, oval then campanulate, undulate, split lengthwise, at first striate, entirely sulcate, in early stages covered with small glassy, micalike, fugacious scales. Gills white or very light brown, then brown, grayish brown with age, eventually soot-brown, adnexed, deliquescent. Stipe 5–10 × 0.3–0.5 cm, white or whitish, cylindrical, silky, often curved, fibrillose then smooth. Flesh thin, pale ochre. Barely detectable odor and flavor. Spores blackish brown, elliptical, smooth, 7.5–10 × 4.5–6 microns.
Edibility Good when young.
Habitat Tufted, on broadleaf stumps and rotting wood, often in the ground.
Season Spring to the first winter frost.
Note One of the most common urban and suburban fungi in North America, it appears in great numbers about stumps or on hidden roots of former street, yard, and park trees.

227 COPRINUS NIVEUS

Etymology   From Latin, "snow-white."
Description   Cap 1.5–5 cm, oval, ogival, quickly campanulate, finally with recurved margin, split and curled over on itself, completely white, floccose and mealy, often squamose. Gills white, then flesh-colored, finally blackish, adnexed, crowded, deliquescent. Stipe 2.5–7.5 × 0.3–0.6 cm, narrowing toward top, densely covered with fugacious, pointed, floccose elements, finally smooth, always completely white, hollow. Flesh white, thin. Undefined odor and flavor. Spores black, broadly elliptical, smooth, 12–18 × 10–13 × 8–10 microns.
Edibility   Of no interest because of size.
Habitat   On dung, especially horse droppings, sometimes also on wet or fertilized ground.
Season   Spring to winter.
Note   *C. roseotinctus* and *C. cothurnatus* are two species which are also floccose and mealy, initially white then respectively pink on a grayish brown background and reddish on a white background. The first grows on the ground beneath broadleaf species, the second on cow dung.

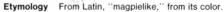

228 COPRINUS PICACEUS

Etymology   From Latin, "magpielike," from its color.
Description   Cap 5–10 cm, soot-black, then variegated with broad, unequal, superficial scales (originating from the break-up of the universal veil) that are white, fugacious and easily detachable, striate lengthwise, sulcate, shiny, smooth and viscid. Gills white then pink, finally black, deliquescent, free, ventricose, up to 1.2 cm long. Stipe 10–25 × 0.6–1.2 cm, white, narrowing toward the top, starting from the basal bulb, fragile, hollow, smooth. Flesh whitish, brown beneath the cuticle of the cap. Odor either nonexistent or nauseous, flavor mucous. Spores blackish and elliptical, 14–18 × 10–12 microns.
Edibility   Not recommended. Probably slightly poisonous, and unpleasant to eat.
Habitat   On the humus in broadleaf woods.
Season   Summer and autumn.
Note   Although this species is frequently reported in North America, there is little evidence that it occurs except along the Pacific Coast from Washington to California.

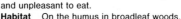

229 COPRINUS PLICATILIS

Etymology From Latin, "wrinkled."
Description Cap 1–3 cm, gray-brown, then ash-gray-blue, darker at center, first oval-cylindrical then campanulate, finally expanded, sulcate-plicate radially, almost diaphanous, glabrous. Gills cream-colored then gray, eventually blackish gray, separated from stipe by a collarium. Stipe 2.5–7.5 × 0.1–0.2 cm, pale, sometimes transparent, cylindrical, smooth. Flesh whitish, extremely thin. No odor or flavor. Spores black, broadly elliptical, smooth, 10–12 × 8–9 microns.
Edibility Of no interest because of size.
Habitat Solitary in grass, meadows, gardens, and by roadsides.
Season Spring to the first winter frosts.
Note This small fungus is easily recognizable by its deeply radially sulcate cap with no ornamentation and with a conspicuous collarium. *Coprophilous C. miser* is quite distinctive when young, with its red-brown or orange-red coloring. *C. auricomus* grows on the ground: the cap is covered with yellowish hairs, most conspicuous on young specimens, and it lacks a collarium.

230 COPRINUS RADIANS

Etymology From Latin, "radiating," because of the mycelium at base of stipe.
Description Cap 2–3 cm, ochreous then paler, membranous, ovate then campanulate, finally expanded, with small, micalike scales, granular at the center, sulcate, margin striate. Gills white, violet-black when mature, adnexed, straight. Stipe 2–4 × 0.4–0.8 cm, white, slightly narrowing toward the top from the enlarged base which originates from a dense mass of orange-brown mycelium. Flesh thin, pale ochre. Odor and flavor negligible. Spores black, elliptical-fusiform, smooth, 9–10 × 4–5 microns.
Edibility Of no interest because of size.
Habitat Often tufted, on trunks, especially elm, or on stacked timber.
Season All year.
Note This is an easily identifiable lignicolous species, even with no carpophore, because of the bright orange-brown tufts of mycelium.

231 COPRINUS SILVATICUS

Etymology   From Latin, "of woods."
Description   Cap 2.5–4.5 cm, ovoid to conical-convex, expanded with recurved margin, leather-ochre with center orange-ochre or cinnamon, then often grayish from margin, sometimes with ochreous disc, apparently smooth, finely pubescent under a lens, deeply plicate-striate or grooved from center, quite often incised at margin. Gills whitish then grayish, finally blackish with some deliquescence, free. Stipe 4–8 × 0.2–0.6 cm, white then ochreous especially in lower part, striate with silky fibrils, pruinous and pubescent at top, quite fragile. Flesh thin, almost nonexistent in cap, whitish in stipe, ochreous at base. No particular odor or flavor. Spores blackish brown, almond- or lemon-shaped, smooth, 11–15 × 6–10 microns.

Edibility   Of no interest because of size.
Habitat   Usually on rotting wood, quite often on the ground or on leaf litter, possibly associated by the mycelium with buried wood, tufted.
Season   Spring to the first winter frost.
Note   This is one of a large number of *Coprinus* species that look more or less alike in the field.

---

232 GYROPHRAGMIUM DUNALII

Etymology   After the mycologist Dunal.
Description   Cap 3–3.5 cm, dry, smooth, ochreous white, grayish, or marked by violet-black spores, fairly irregularly cracked, papyruslike. Gills (pseudogills) violet-black, formed by radially placed, triangular tubercles, 4–6 per line, wrinkled, at first soft then dry and very fragile. Stipe 8–10.5 × 0.6–1.5 cm, narrowing at both ends, lower part with peridial remains (fragments of varying sizes and shapes), sometimes like one or more rings, sometimes volvalike, with yellow markings at base. Pale straw-colored flesh, compact but light, suberose. Slight cyanic odor and flavor. Spores blackish brown, globose to elliptical, smooth, 4.5–6.5 microns.

Edibility   Of no interest because of size.
Habitat   In sandy places, coastal or desert dunes, worldwide, but rare.
Season   Autumn to spring.
Note   This strange Gasteromycete resembles a dry field mushroom. This is but one of a large variety of desert puffball-like fungi that, unlike gilled mushrooms, do not forcibly eject their spores. Rather, the spores are retained within the fruit-body and kept from exposure to extreme heat and evaporation until the rainy season.

233 CHROOGOMPHUS RUTILUS

Synonym *Gomphidius viscidus.*
Etymology From Latin, "yellowish red."
Description Cap 3–9 cm, viscid but soon drying, copper- or wine-colored, often brick-red or brown, margin recurved. Gills large, distant, sometimes forked, decurrent, easily detachable, red then blackish. Stipe 4–8 × 0.8–1.5 cm, yellow then cap-colored, long, enlarged toward top, full, first viscid then with squamose ringlike zones, then with velar remains just beneath gills. Flesh yellowish, darker in stipe, yellower at base, soft. Mild odor, pleasant flavor. Spores blackish brown, elongated-fusiform, smooth, 16–22 × 6–8 microns.
Edibility Mediocre; remove the cuticle.
Habitat In groups in coniferous woods, especially beneath pines, in moss.
Season Late spring to late autumn.
Note The genus *Chroogomphus* has been separated from the genus *Gomphidius* because: it has a fibrillose veil which dries quickly and is not glutinous; the flesh of the cap is colored and, along with the basal mycelium, stains violet in an iodine solution (Melzer's Reagent). All species grow under conifers, and most are western in distribution.

234 GOMPHIDIUS MACULATUS

Etymology From Latin, "spotted."
Description Cap 3–5 cm, fleshy, convex then open, fairly depressed, almost invariably obtuse with no umbo, somewhat viscid, often with black spots, margin involute. Gills whitish then gray, finally blackish, distant, decurrent, ventricose, detachable. Stipe 6–7 × 0.8–1 cm, white, speckled with small blackish or pale purple spots or fibrils, yellow at base, not viscous and without a veil. Spores olive-black, fusiform, smooth, 20–30 × 7–9 microns.
Edibility Good, but remove the cuticle.
Habitat In groups in mountains beneath larch.
Season Summer and autumn.
Note In the more slender forms some distinguish a separate species, *G. gracilis.* One distinguishing feature is the fact that the flesh darkens to black without turning reddish first. *G. subroseus* is easily identifiable by its bright pink-red viscid cap contrasting sharply with stipe and gills which remain white for a long time. Stipe has a yellowish base and is sheathed, like most other *Gomphidius* members, by a glutinous veil that forms a sort of ring. Like *Chroogomphus*, species of *Gomphidius* grow under conifers, and are most common across northern North America.

235 GOMPHIDIUS GLUTINOSUS

Etymology From Latin, "viscid" or "slime-covered."
Description Cap 5–12 cm, fleshy, glabrous, very viscous, violet-brown, often with black spots, cuticle completely detachable. Gills whitish then soot-colored, detachable, distant, decurrent. Stipe 6–12 × 1.2–2.5 cm, white at top, yellow at base, solid, soft, with fugacious viscous veil. Flesh white and soft in cap, pale ash-colored beneath cuticle, tough in stipe, yellow at base. Spores blackish, elongated-fusiform, smooth, 18–22 × 6–8 microns.
Edibility Good, but remove the glutinous cuticle.
Habitat Several to many in coniferous woods.
Season Summer and autumn.
Note G. glutinosus is the most common and most widely distributed species in the Gomphidius-Chroogomphus group. Its nearest look-alike is G. oregonensis, a western species that is usually cleaner in appearance and grows in cespitose clusters.

236 PAXILLUS INVOLUTUS

Etymology From Latin, "curly-edged."
Description Cap 7–20 cm, rust-ochre, fleshy, flat-convex then depressed, pubescent then smooth, sometimes slightly viscid particularly at center in wet weather, silky when dry, margin obtuse, villose, conspicuously involute for long periods, finally often ribbed and acute when completely expanded. Gills pale ochre then rust-colored, becoming marked with dark brown when touched, decurrent, quite broad, ramified toward stipe where often anastomosed thus forming a reticulum, easily detachable from flesh. Stipe 5–8 × 1–4 cm, dark yellowish, often spotted, enlarged toward top, glabrous or barely pruinose, solid. Flesh reddish yellow then yellow, soft, juicy. Slightly acid and fruity odor, similar flavor. Spores rust-ochre, elliptical, smooth, 8–10 × 5–6 microns.
Edibility Not recommended although tasty because it is now known to cause a gradually acquired hypersensitivity that can result in massive hemolysis and be life-threatening.
Habitat In wet woodland, particularly broadleaf.
Season Early summer to late autumn.
Note P. vernalis has a chocolate-brown spore print and grows near aspen.
Caution See edibility.

237　PAXILLUS ATROTOMENTOSUS

Etymology　From Latin, "black and very velvety."
Description　Cap 5–30 cm, convex, slightly irregular and humped then expanded and depressed, quite often spatula- or kidney-shaped, cuticle entirely detachable, thick, reddish yellow-brown, olive-brown, dry, velvety, finally almost glabrous or finely areolate, margin often curled, whitish yellow then cap-colored. Gills crowded, ramified and anastomosed, often alveolate toward stipe, arcuate, decurrent, soft, fragile, easily detachable, cream-colored then yellow-ochre, yellow tinged with orange, brownish when touched, edge darkening with age. Stipe central or eccentric or lateral, 4–9 × 2.5–5 cm, sturdy, solid, rooting, covered to top with down ranging from dark brown to blackish brown. Flesh thick, soft, watery, pale yellow with pale lilac-pink speckling or marbling in cap and brownish toward base. Slightly fetid odor, very bitter flavor. Spores rust-brown, elliptical, smooth, 6–7 × 3–4 microns.
Edibility　Extremely bitter even after boiling, possibly poisonous in the manner of P. involutus (**236**).
Habitat　On old stumps and dead roots of conifers.
Season　Summer and autumn.
Caution　Possibly poisonous.

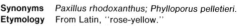

238　PHYLLOPORUS RHODOXANTHUS

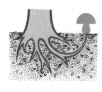

Synonyms　*Paxillus rhodoxanthus; Phylloporus pelletieri.*
Etymology　From Latin, "rose-yellow."
Description　Cap 2–10 cm, orange-brown, olive-brown toward margin, hemispherical then flat, sometimes depressed, without umbo, surface uniformly velvety, margin interrupted. Gills bright golden yellow, interconnected by septi, anastomosed, alveolate, distant, large and long, adnate, decurrent, detachable. Stipe 2–6 × 0.5–1.5 cm, first reddish yellow then brownish, pale green-yellow at base, cylindrical, almost rooting at base, often arcuate, solid, fibrillose. Flesh yellowish, wine-red beneath cuticle and in stipe, color intensifying on exposure to air, thick at center, soft. Odor pleasant, flavor sweet. Spores olive-yellow, elliptical-fusiform, smooth, 10–12.5 × 4–4.5 microns.
Edibility　Good.
Habitat　Single to many in broadleaf and coniferous woods, usually in acid soil.
Season　Summer and late autumn.
Note　This is a strange fungus which acts as a link between the Boletaceae with gills (*Paxillus, Gomphidius*) and those with pores (*Boletus*).

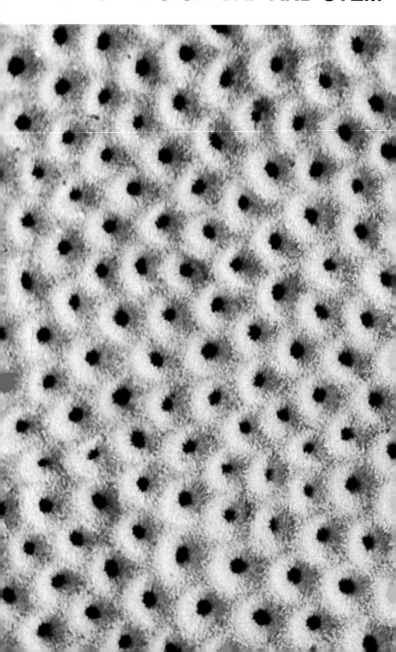

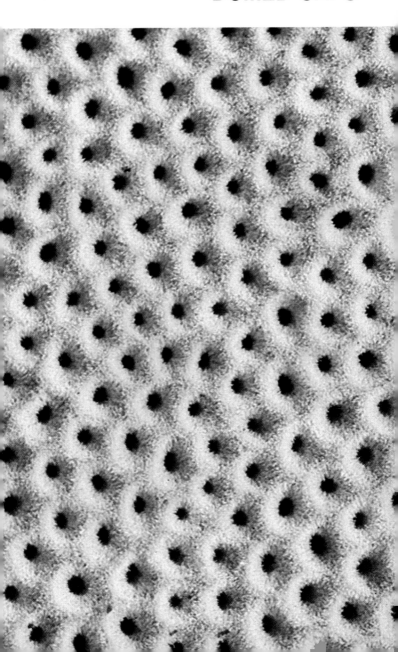

239 BOLETUS AEREUS

Etymology   From Latin, "bronze," from the cap color.
Description   Cap 10–30 cm, hemispherical then convex, cuticle finely velvety, dry, never smooth, dark brown, sometimes almost blackish or brown tinged with rusty color, edge often lobate. Tubes first whitish, pale green with age, pores initially white covered with fine pruinescence, then tube-colored. Stipe 7–15 × 3–6 cm, solid, massive, club-shaped when mature, with brownish reticulum that normally does not reach base. Flesh soft, unchanging, white. Nice odor and flavor. Spores olive-brown, fusiform, smooth, 12–16 × 4–5 microns.
Edibility   Excellent, with firm flesh.
Habitat   In European oak and chestnut woods, in open or sunny places.
Season   Summer and autumn.
Note   The closest North American look-alike is *B. variipes,* with a soft, dry, wrinkled buff-gray to ochre cap, growing in open oak woods in summer. It is also an excellent edible.

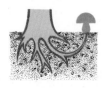

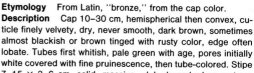

240 BOLETUS EDULIS

Common names   Cep; edible bolete.
Etymology   From Latin, "edible."
Description   Cap 5–25 cm, hemispherical, convex then flattened, cuticle smooth, slightly viscous in damp weather, barely detachable, whitish, ochreous, light brown, or chestnut, not uniform. Fairly long tubes, almost free from stipe, first whitish then greenish yellow; pores small, circular, tube-colored. Stipe 4–18 × 2–5 cm, solid, bulging or cylindrical, white or light ochre, covered by a reticulum first white then slightly darker than background. Flesh white, unchanging, soft then softening further. Odor nice, flavor sweetish, tasty, like hazelnuts. Spores olive-brown, fusiform, smooth, 14–17 × 4.5–6.5 microns.
Edibility   Excellent.
Habitat   Beneath conifers and broadleaf trees.
Season   In damp, cool conditions, in spring, reappearing from late summer until first cold spell.
Note   There are many recognized varieties of this species; key characteristics are the slightly viscid cap, the white reticulation on the young stipe, and its pleasing odor and taste. It is sold dried in gourmet shops.

241 TYLOPILUS FELLEUS

Synonym *Boletus felleus.*
Etymology From Latin, "of gall," from its bitterness.
Description Cap 5–15 cm, hemispherical then convex, finally flat, cuticle dry, finely velvety, difficult to detach, light brown. Tubes long, semifree, increasing in size toward stipe and margin, longer at center, whitish or cream-colored then pinkish; pores round, quite small, tube-colored. Stipe 4–12 × 1–4 cm, solid, slightly club-shaped, almost cap-colored, slightly lighter tinged with olive, marked dark brown reticulum, surface velvety toward base. Flesh soft, white, faintly pinkish when exposed to air, slightly brown beneath cuticle. Odor insignificant, flavor very bitter. Spores pinkish, elliptical or fusiform, smooth, 9–18 × 4–6 microns.
Edibility Much too bitter to eat.
Habitat On fine, acid or acidified soil, beneath coniferous and broadleaf species, often in rotting trunks.
Season Summer and autumn.
Note This is not infrequently mistaken for *Boletus edulis* **(240)**, because of its shape, but it is bitter, the reticulation on stipe is dark, and pores turn pinkish. *T. indecisus* is nearly identical but mild in flavor and edible. *Tylopilus*, in general, has pink pores and a pinkish spore print, and often a bitterish taste.

---

242 BOLETUS LURIDUS

Etymology From Latin, "dirty" or "filthy" because of the colors it turns when touched or cut.
Description Cap 5–12 cm, hemispherical then convex, cuticle dry, velvety in young specimens, smooth with age, sometimes viscid in damp weather, color variable from leather- to olive-brown, or chamois to rust-brown, even on the same carpophore, darkens when touched. Tubes free, long, yellow to greenish, turning blue when touched; pores small, roundish, reddish to orange, bruising greenish blue. Stipe 4–20 × 1.5–5 cm, soft, solid, bulging then club-shaped or cylindrical, reddish yellow, largely covered by a conspicuous, elongated red reticulum, with bright red pruinescence in upper part, surface turning blue-green when exposed, may be red at base of stipe if invaded by larvae, with a distinctive orange-red coloration at tube attachment. Odor fruity, flavor sweetish. Spores olive-brown, ovoid or almond-shaped, smooth, 11–15 × 5–8 microns.
Edibility Eaten in Europe; **not recommended in North America** because of the difficulty of accurate identification, and reports of vomiting and diarrhea from eating this or related red- to orange-pored species.
Habitat Beneath broadleaf trees, especially oak.
Season Spring to autumn.
Caution Can cause vomiting and diarrhea.

243 BOLETUS PARASITICUS

Etymology From Greek, "parasitic," from its habitat.
Description Cap 3–8 cm, hemispherical then convex, and finally flattened, cuticle downy, tending to crack in dry weather, not detachable, color light ochre, sometimes tinged olive. Tubes short, slightly decurrent, yellow, slightly olive-brown when mature; pores large, round to polygonal, tube-colored and also spotted with rust-brown. Stipe 2.5–7 × 0.5–1.5 cm, solid, slender, tapering and curved at base, yellowish, fibrillose, speckled with brown floccose elements at top. Flesh soft, quite leathery, rarely rots but dries out naturally, yellow, sometimes turning blue at tube attachments and at edges of stipe. Odor insignificant, flavor sweetish. Spores olive-yellow, cylindrical or fusiform, smooth, 10–17 × 4–6 microns.
Edibility Good.
Habitat Isolated or in groups, parasitic on carpophores of *Scleroderma*.
Season Summer and early autumn.
Note A unique bolete which grows only on *Scleroderma*, particularly *S. citrinum* (*S. aurantium*). The size of the host carpophores affects the size of the bolete.

244 BOLETUS QUELETII

Etymology After the French mycologist Quélet.
Description Cap 5–20 cm, hemispherical then convex, cuticle slightly velvety, orange to chrome-yellow to ochre, large in young specimens, when old reduced in such a way that the hymenium juts out. Tubes semifree, yellow to pale green-yellow turning blue; pores small, round, orange, tinged with pale purple-red, lighter in color, yellow toward edge, turning blue. Stipe 5–18 × 1.5–4 cm, solid, soft, fairly club-shaped or fusiform with base slightly rooting, beet red at base, tinged with yellow toward top, sometimes with light green coloration halfway up. Flesh soft, yellow, red in stipe, turning blue when exposed to air then back to slightly grayer original colors. Odor fruity, slightly acid, flavor sweet with slightly bitter aftertaste. Spores olive-brown, fusiform, smooth, 8–17 × 5–7 microns.
Edibility Reportedly eaten in Europe but **not recommended.**
Habitat Prefers under broadleaf species in open areas.
Season Late spring to autumn.
Note This species is not known in North America. One similar species, growing under broadleaf trees, is the poisonous *B. subvelutipes*, with dark red hairs at the base of the stipe, and flesh bruising blue to blue-black.

245   BOLETUS SATANAS

Etymology From Hebrew, "evil."

Description Cap 6–40 cm, subglobose then fairly convex, margin undulate, cuticle finely velvety, then smooth and, in dry weather, cracked, whitish tinged ash-gray or greenish, then light brown. Tubes shorter toward stipe and edge, longer in center, yellowish, pale blue when touched; pores small, slightly irregular, yellow then orange-red, turning blue. Stipe 4–20 × 3–10 cm, solid, squat, sturdy, fat, reticulum with polygonal elements varying in color from bottom to top and from fungus to fungus, brown at bottom, then reddish in middle, pink-red or white in upper part, on a yellow background at top, red lower down, turning blue. Flesh soft, then softening further, quick to rot, white, pale yellow at tube attachments, slightly spotted with pink to violet to blue when exposed. Strong odor of cabbage or dung when mature, flavor initially mild, walnutlike then nauseous. Spores olive-brown, ovoid, smooth, 11–15 × 5–7 microns.

Edibility Poisonous, as are some other red-pored boletes.

Habitat In small groups beneath broadleaf trees, especially oak; reported along coastal California.

Season Spring to autumn.

Note A few boletes are very good edibles; most are edible but undistinguished; a few (mostly those that stain blue and have reddish pores) are poisonous.

---

246   BOLETUS ZELLERI

Etymology After the American mycologist Zeller.

Description Cap 5–10 cm, convex to flat with dry cuticle, with white pruinescence when young, then downy, sometimes areolate, dark brown, margin often reddish. Tubes quite long, adnate, initially olive-yellow then deep yellow, turning to blue when touched; pores polygonal, tube-colored. Stipe 5–8 × 0.7–1.3 cm, solid, cylindrical or tapering toward the top, with whitish mycelium at the base, fairly finely spotted or uniformly colored with red, on yellow background toward the base. Flesh in stipe in young specimens yellow, in adult specimens reddish, pale yellow in cap, turning blue in some parts. Odor insignificant, flavor slightly acidic. Spores olive-brown, ellipsoid, smooth, 12–15 × 4–5 microns.

Edibility Edible but not choice.

Habitat Solitary or in small groups, in forests on the West Coast.

Season Spring.

Note This species is quite similar to *B. chrysenteron* (**248**), differing mainly in its darker-colored pruinose cap and in the microscopic structure of the cuticle.

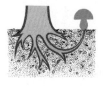

247 GYROPORAS CASTANEUS

Synonym *Boletus castaneus.*
Etymology From Byzantine Greek, "chestnut."
Description Cap 3–10 cm, hemispherical to convex, sometimes depressed, sometimes irregular, cuticle detachable from
edge, thin, dry, at first feltlike-velvety then almost smooth, reddish brown, fairly deeply colored. Tubes free and always
shorter toward stipe, whitish and detachable; pores roundish,
white then lemon-yellow, darkening when bruised. Stipe 4–7 ×
1–3 cm, easily detachable from cap, sinuate toward top and
base, often slightly curved, "knotty," pithy, cavernous then
hollow, fragile, cap-colored. Flesh fragile, white, sometimes
slightly pinkish when exposed, brownish beneath cuticle and
on sides of stipe. Pleasant odor, hazelnutlike flavor. Spores
light lemon-yellow, elliptical, smooth, 8.5–12.5 × 5.5–6.5 microns.
Edibility Very good when young.
Habitat Singly and in small groups, beneath broadleaf trees,
especially oak.
Season Summer and autumn.
Note This mushroom is best to eat while the pores are still
white. *Gyroporus* is distinguished as a genus by its yellowish
spores and the stipe, which is hollow or becomes cavernous.

248 BOLETUS CHRYSENTERON

Etymology From Latin, "with a golden interior."
Description Cap 3–12 cm, hemispherical then convex, finally flat, various shades of brown, cuticle dry, velvety, easily
cracking revealing slight purple-pink subcuticle stratum. Tubes
yellow then pale green, turning blue when touched; fairly large,
polygonal, tube-colored pores. Stipe 3–10 × 0.5–2 cm, solid,
cylindrical, tapering at base, slightly curved, with striations in
upper part that extend into hymenium, yellowish tinged with
brown or pale red, turning blue, yellow mycelium at base. Flesh
tender, soft when old, pale yellow turning blue and then pinkish, purple-pink under cuticle. Odor and flavor pleasant.
Spores olive-brown, fusiform, smooth, 11.5–15.5 × 4–7 microns.
Edibility Fair to good.
Habitat Beneath broadleaf or coniferous trees.
Season Late spring to late autumn.
Note *B. chrysenteron* is a complex of closely related species
that often can only be reliably differentiated by microscopic examination. The tendency to turn blue or green on bruising is
sometimes delayed or partial.

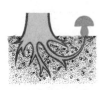

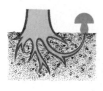

249 PORPHYRELLUS PSEUDOSCABER

Synonyms *Tylopilus pseudoscaber; Boletus pseudoscaber.*
Description Cap 5–16 cm, hemispherical then convex, finally flat, cuticle velvety in dry weather, smooth when wet, very dark brown, bruising blue then purplish brown. Tubes long, semifree, dark brown to purple-brown, bruising blue; pores small, angular, tube-colored, blue then blackish when touched. Stipe cylindrical, 4–16 × 1–3 cm, slightly club-shaped, solid, hard, almost cap-colored, outer surface pubescent, sometimes with striations that may resemble a false reticulum. Flesh firm when young, then soft, fibrous in stipe, whitish, turning pink and then gray when exposed, becoming pale blue at tube attachments. Faint, though slightly acrid odor and flavor. Spores purple-red-brown in mass, fusiform, elliptical, smooth, 10–20 × 6–10 microns.
Edibility Mediocre.
Habitat Solitary or in small groups under conifers.
Season Summer and autumn.
Note This is the most common species of *Porphyrellus,* and is found throughout the coastal forests of the Pacific Northwest. Species of *Porphyrellus* can only be differentiated microscopically. They are widely distributed across northern North America.

---

250 SUILLUS CAVIPES

Synonyms *Boletinus cavipes; Boletus cavipes.*
Etymology From Latin, "hollow-legged."
Description Cap 5–20 cm, hemispherical or conical, finally slightly concave, chestnut-colored, reddish brown, initially with white margin, partial velar remains, cuticle with tufts of brown hairs, dry. Tubes short, very adherent to flesh, yellow to pale green; pores elongated, rhomboid, polygonal, from yellow to pale green. Stipe 5–8 × 0.5–2.5 cm, hollow, cylindrical, slightly club-shaped, slightly curved, cap-colored but lighter, ring whitish, floccose. Flesh quite tender, pale yellow, unchanging. Odor insignificant, flavor sweet. Spores olive-yellow, fusiform, smooth, 7–15 × 3–4 microns.
Edibility Mediocre.
Habitat In small groups beneath larch.
Season Late summer and autumn.
Note The dry, scaly cap, hollow stipe, and its occurrence under larch make this species easy to recognize. A similar mushroom but with a solid stipe, common under eastern white pine, is *S. pictus;* another look-alike, common under Douglas fir in the West, is *S. lakei.*

251 SUILLUS GREVILLEI

Synonym *Boletus elegans.*
Etymology After the Scottish mycologist Greville.
Description Cap 5–12 cm, lemon-yellow or golden yellow then reddish orange, conical-obtuse then convex-flattened, sometimes obtuse and umbonate, cuticle detachable, viscid in wet weather, fleshy, soft, margin incurved, regular, lighter than disc. Tubes yellow then rose-gray, adnate, decurrent, fine, detachable, short; pores round then polygonal, tube-colored, sometimes marked with reddish purple, thin. Stipe 5–10 × 1.5–2.5 cm, yellow, orange-red, with brownish reticulation above ring, brownish striations or speckling below ring, blackish brown or olive-colored at base, fleshy, solid; ring yellowish white, quite viscid, ascending. Flesh yellow in cap, slightly greenish at base of stipe, soft but soon limp, fibrous in stipe. Faint geranium odor, flavor sweet. Spores olive-yellow, ellipsoid-fusoid, smooth, 7–11 × 2–4 microns.

Edibility Fair if the cuticle and ring are removed.
Habitat In groups under larch, in sunny, grassy areas.
Season Late spring to autumn.
Note Also growing only under larch are *S. cavipes* (**250**); *Fuscoboletinus ochraceoroseus*, with dry, scaly, rose-red cap and wine-brown spores; and *F. spectabilis*, with slimy, streaked to scaly reddish cap.

252 SUILLUS GRANULATUS

Synonym *Boletus granulatus.*
Etymology From Latin, "dotted."
Description Cap 4–18 cm, hemispherical, convex then flattened, cuticle viscous in damp weather, smooth and shiny when dry, entirely detachable, yellow tinged with brown or reddish. Tubes first short, lengthening with age, adherent to stipe, pale yellow then yellow with pale green shades; pores small, angular, emitting droplets of opalescent latex when young, first markedly yellow then pale green-yellow. Stipe 4–10 × 1–2 cm, cylindrical, often curved, solid, light yellow then marked with brown, upper part with small reddish brown markings caused by dried drops of latex, present as such in young specimens. Flesh tender, soft and spongy when old, pale yellow or white, unchanging. Pleasant, slightly acid, fruity odor, flavor sweetish. Spores brown-ochre, fusiform, smooth, 8–10 × 3–4.5 microns.

Edibility Good without cuticle, or dried.
Habitat In groups, beneath conifers, especially pine.
Note Several common boletes have slimy cap cuticles, yellowish pores, and stipes with dark dots; most grow under pines; and all are edible and tasty once the cuticles have been removed.

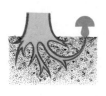

253 SUILLUS LUTEUS

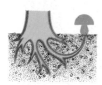

Synonym *Boletus luteus.*
Common name Slippery Jack.
Etymology From Latin, "yellow."
Description Cap 4–18 cm, hemispherical or somewhat coni-cal then convex with a slight umbo at times, margin regular with some partial velar remains, various shades of brown, violet highlights in damp weather, tending to fade with age and dry weather, cuticle viscous and detachable. Tubes adherent or briefly decurrent on stipe, sulfur-yellow then slightly pale green; pores tiny even when mature, round, tending to become polygonal, tube-colored but darkening slightly with age. Stipe 3–13 × 1–3 cm, cylindrical, solid and soft, whitish or pale yel-low, with pale yellow dots which turn reddish brown, brownish at base, ring membranous, sleevelike, whitish, pale purple be-neath. Flesh tender, soft, pale yellow, whitish, unchanging. Slight odor and sweet flavor. Spores brown-ochre, fusiform, smooth, 7–10 × 3–3.5 microns.
Edibility Good after removing cuticle and drying, although some experience diarrhea.
Habitat Gregarious beneath conifers, preferably pine.
Season Late spring to late autumn.

254 SUILLUS PLACIDUS

Synonym *Boletus placidus.*
Etymology From Latin, "peaceful" or "placid."
Description Cap 3–13 cm, hemispherical, convex, finally flattened, sometimes depressed, cuticle viscous, detachable, white to yellow tinged with pink, darkening with age to brown from the center. Tubes slightly decurrent, not easily detachable in young specimens, whitish to yellow tinged with greenish; pores small, irregular, with drops of opalescent latex that on drying produce small spots varying from orange-yellow to red-dish-brown, at first tube-colored, then speckled reddish brown or violet. Stipe 5–15 × 0.5–3 cm, cylindrical, sinuate, some-times slightly eccentric, solid, with no ring, pale yellow at top, brown or pinkish lower down, where the droplets of dried latex form a purple ornamentation. Flesh soft, white, pale yellow at edges, violet-gray when exposed to air. Odor insignificant, fla-vor sweetish. Spores dark ochre, elliptical or fusiform, smooth, 7–9 × 3–3.5 microns.
Edibility Fair after removing cuticle.
Habitat Several to many, under eastern white pine.
Season Summer and autumn.
Note The smooth, slimy white cap, mild taste, lack of stipe ring, and its occurrence under white pine make this an easy bolete to identify.

255 BOLETUS PULVERULENTUS

Etymology From Latin, "dusty."

Description Cap 3–15 cm, hemispherical, convex, rarely flat, cuticle velvety when young, when mature quite smooth or viscid in wet weather, not easily detachable, Havana-brown, olive-ochre-gray or brownish, quickly turning dark blue when touched, then blackish. Tubes long, fine, adnate, yellow to greenish, turning blue; pores quite large when mature, polygonal, tube-colored. Stipe 4–10 × 1–2 cm, solid, tapering, rooting toward base, reddish brown, pale purple on ochre background, darker at base, yellow at top, fairly uniform brown when mature, turning dark blue then blackish, initially velvety. Flesh tender then soft, bright yellow, when exposed to air immediately dark blue then sky-blue-gray. Odor fruity, flavor slightly acid, sweetish. Spores olive-colored, fusiform, smooth, 11–15 × 4–6 microns.

Edibility Fair when cooked.

Habitat Beneath broadleaf trees, also rarely beneath fir, on acid ground, in damp places.

Season Late spring to early autumn.

Note Its typically dark brownish cap, large pores, and very quick color change to blue, make this a distinctive species. Although not common, it is found throughout eastern North America, west to Michigan.

---

256 SUILLUS VARIEGATUS

Synonym Boletus variegatus.

Etymology From Latin, "variegated" or "many-colored."

Description Cap 5–15 cm, hemispherical then convex, finally flattened, cuticle only slightly viscid in wet weather, covered with brown hairy tufts which may drop off with age, standing out against pale straw-yellow or ochreous background. Tubes adnate or slightly decurrent, pale yellow then olive-brown, blue when touched; pores roundish to angular, tube-colored. Stipe 3–13 × 1.5–3 cm, solid, cylindrical, sometimes slightly enlarged at base, slightly velvety in lower part, cap-colored. Flesh tender, soft, watery in wet weather, pale yellow, slightly orange, reddish-brown at base, turning slightly blue. Odor slightly acid, flavor quite pleasant, sweetish. Spores olive-brown, elliptical or fusiform, smooth, 7.5–10.5 × 3–4 microns.

Edibility Mediocre; the flesh turns gray when cooked.

Habitat Gregarious in pinewoods.

Season Summer and autumn.

Note A close look-alike found under 2-needle pines from Michigan west is S. tomentosus; yellowish to yellowish orange, it has a fibrous-scaly cap and its flesh turns bluish to greenish when exposed to air.

257 LECCINUM AURANTIACUM

Synonym *Boletus aurantiacus.*
Etymology From Latin, "golden orange."
Description Cap 4–25 cm, subspherical to fairly convex, cuticle with margin which, for a short section, sheaths the stipe in yet-unopened specimens, orange-red, various shading, tending to discolor, finely velvety. Tubes long, thin, free, easily detachable, white or cream then grayish; pores small, round, tube-colored. Stipe 8–25 × 1.5–5 cm, solid, soft, elongatedly club-shaped, with small raised scales, whitish then brown to black, clearly standing out against whitish background. Flesh quite soft in cap, fibrous in stipe, white turning reddish brown then blackish gray when exposed. Odor pleasant, flavor sweetish. Spores light Havana-brown, fusiform, smooth, 12–20 × 4.5–5 microns.
Edibility Good; darkens when cooked.
Habitat Scattered beneath pines and aspen.
Season Summer and autumn.
Note This complex of microscopically differing species is distinguishable from the *L. insigne* complex by color changes in the cut flesh. Members of the former turn reddish then grayish, whereas members of the latter turn lilac-gray and then darken.

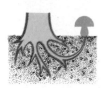

258 LECCINUM GRISEUM

Synonyms *Leccinum carpini.*
Etymology From Latin, "gray cap."
Description Cap 4–15 cm, hemispherical then campanulate, gray to ochreous to dark brown, cuticle wrinkled, finely pruinous when young, then glabrous, tending to split or crack, slightly viscid when wet. Tubes very long, free, straw-colored when young, grayish when mature; pores small, round, tube-colored. Stipe 5–20 × 1–3 cm, slender, cylindrical, or rather club-shaped, solid, fibrous, surface rough, covered with whitish scales that blacken, standing out against the grayish white background. Flesh tender then soft in cap, fibrous in stipe, white, pale yellow at tube attachments, turning violet-gray on exposure to air, gradually darkening. Faint, pleasant odor and flavor. Spores cigar-brown, fusiform, smooth, 11–20 × 5–7 microns.
Edibility Mediocre, best in sauce; black when dried.
Habitat Fairly gregarious, beneath broadleaf trees, especially birch, poplar, hornbeam, and oak.
Season Summer and autumn.
Note Of the many boletes with gray-brown caps and scabrous stipes, the most common in North America is *L. scabrum;* the differences are only at the microscopic level.

259 BOLETOPSIS SUBSQUAMOSA

Synonyms *Polyporus leucomelas; Polyporus griseus.*
Etymology From Latin, "somewhat scaly."
Description Cap 4–17 cm, hemispherical to convex, irregularly humped and lobate, gray to olive-gray to blackish gray where damaged, fibrillose, tending to crack. Tubes short, decurrent, whitish then brown; pores quite broad, irregular, tube-colored. Stipe 2–8 × 1–3 cm, sturdy, solid, cap-colored, shape variable. Flesh white, when exposed to air rose then gray, firm. Odor insignificant, flavor bitterish. Spores whitish, oval and irregularly angular, with protruding nodes, 6–4 × 3.5–4.5 microns.
Edibility Palatable if soaked, parboiled, and then cooked.
Habitat On acid ground beneath broadleaf and coniferous trees.
Season Summer and autumn.
Note This looks like a bolete but has very short tubes and tuberculate spores. Although several species are sometimes recognized, based on cap color, there are no significant microscopic differences, and this is now seen as a monotypic species with lighter and darker forms.

260 POLYPORUS TUBERASTER

Synonym Canadian tuckahoe.
Etymology From Latin, "tuberous."
Description Cap 5–20 cm, flat to funnel-shaped, yellowish ochre, covered with hairy scales, brown, adpressed, margin at first involute then expanded, fleshy, soft. Tubes white, decurrent; pores polygonal, not broad, white. Stipe 5–20 × 2.5–4 cm, sturdy, solid, quite elastic, full, glabrous, ochreous, brown at base which emerges from large hypogeal sclerotium. Flesh whitish, soft but elastic. Odor and flavor resemble cheese. Spores white, ovoid or elliptical, 12–14 × 3.5–4 microns.
Edibility Good and tender while young, but then needs lengthy cooking.
Habitat On the ground in broadleaf woods, especially under oaks and beech.
Season Spring to late autumn.
Note The mycelium of this fungus forms large, fairly compact sclerotia with the consistency of porous rock. When cut, the blackish brown mass is marbled with white clumps of hyphae. This is sold in European markets, and, if well watered, will produce carpophores at home.

261 ALBATRELLUS OVINUS

Synonym *Polyporus ovinus.*
Etymology From Latin, "pertaining to sheep."
Description Cap 6–10 cm, convex then flat or depressed, irregularly humped, margin sinuate, sometimes lobate, cuticle detachable, dry, thick, whitish turning yellow, faintly brown with age, often cracked, areolate. Tubes short, 1–2 mm, decurrent, white; pores very small, fairly round, whitish then pale yellow. Stipe 3–6 × 1–2 cm, central or eccentric, irregular, solid, hard, whitish sometimes marked with brown, pruinose, often concrescent. Flesh white, slightly yellow at base, thick, firm but fragile. Odor somewhat acid, fruity, flavor sweet, of almonds, sometimes a little bitter. Spores white, ovoid, smooth, 3.5–4.5 × 3 microns.
Edibility Good, especially when preserved in oil.
Habitat In large groups often joined together by cap or stipe, in mountainous regions beneath conifers.
Season Summer and autumn.
Note *A. confluens,* cap pinkish orange-brown, grows in the same locale. The carpophores have a greater tendency to grow with cap and stipe joined. *A. cristatus,* common in southeastern North America, has a greenish yellow cap soon areolate and often concrescent. Only edible when young, it is leathery and astringent at maturity.

262 ALBATRELLUS PES-CAPRAE

Synonym *Polyporus pes-caprae.*
Etymology From Latin, "like a goat's foot."
Description Cap 6–12 cm or more, semicircular or kidney-shaped, sometimes irregularly lobate, convex then flat or depressed toward stipe, cuticle dry, not detachable, feltlike and squamose, covered with scales at first adpressed then raised, often delimited by cracks, reddish brown, margin initially incurved. Pores not detachable, short, decurrent, white or light lemon-yellow, finally beige, wide, polygonal. Stipe 3–5 × 1–2 cm or more, eccentric or lateral, whitish, pale yellow at base, irregular, with brownish reticulum at top, squamose like cap. Flesh white tending to lemon-colored, reddish at base, firm, thick. Faint pleasant odor, flavor of hazelnuts. Spores white, elliptical or oval, smooth, 8–11 × 5.5–7.5 microns.
Edibility Excellent.
Habitat On acid ground, solitary or in small concrescent groups, in broadleaf and coniferous woods.
Season Summer and autumn.
Note Not common in North America, this genus has more than a dozen species of terrestrial polypores that are stalked and resemble boletes, differing, in part, by having very short tubes not detachable from cap flesh.

263 POLYPORUS ARCULARIUS

Synonym *Polyporellus arcularius.*
Etymology From Latin, "vaulted," from pore shape.
Description Cap 2–5 cm, when mature convex then umbilicate, not zoned, brown, grayish-brown, tending to ochreous when drying, first covered with small darker scales, abundant toward margin which normally appears ciliate. Tubes white, adnate-decurrent, not removable; pores white then light brownish ochre, broad, oblong, rhomboid. Stipe 1–2.5 × 0.1–0.2 cm, brownish gray or blackish brown, at first barely squamulose, then glabrous. Flesh white, leathery. Mushroomy odor. Spores white, globose, smooth, 3 microns.
Edibility Inedible.

Habitat On dead branches and sticks in woodland or on the ground growing from buried wood, especially beech, birch, elm, maple, oak, and poplar.
Season Especially in the spring, but into summer.
Note *P. brumalis* is similar, with dark grayish brown, hairy or squamulose cap, margin at most velvety, pores at first round, and stipe velvety or squamulose, not glabrous. It occurs on dead branches, and is often seen over winter and into early spring. The *P. badius* complex is a group of similar species distinguished by the stipe, which is partially or entirely black.

---

264 GANODERMA LUCIDUM

Etymology From Latin, "shiny" or "brilliant."
Description Carpophore 5–28 cm, cap circular or kidney-shaped, covered with a shiny crust, cap zoned from yellow to dark red, margin white or yellow. Stipe, when present, 5–18 × 1–5 cm, usually eccentric-lateral, surface like that of the cap. Tubes white then cinnamon-colored, small and round. Flesh whitish spongy, then reddish and ligneous, zoned. Spores brown, elliptical, finely warty, 10–12 × 6–8 microns.
Edibility Inedible once encrusted, but soft and flavorful when it first appears.

Habitat Typically at the base of living maples and oaks, but also on stumps and roots of these and other broadleaf trees; rarely reported on conifers.
Season Late spring to late autumn.
Note An attractive species when fresh, easy to dry and preserve as a decorative object. In the Orient it is called *ling chi*, and it is cultivated and sold commercially, and used for many of the same reasons people use ginseng: to alleviate minor disorders, increase vitality, and ensure long life.

265 POLYPORUS MORI

Synonyms *Favolus alveolaris; Favolus canadensis; Favolus europeus.*

Description Cap 2–8 cm, roughly kidney-shaped, thin, ochreous cream-colored, cuticle not detachable, marked by radial fibrils, sometimes joined to form small adpressed scales, smooth, tending to fade, margin curved then expanded. Tubes short, decurrent, whitish; pores broad, alveolate, polygonal, often with fine hairs. Stipe 0.4–1 × 0.5–0.8 cm, lateral, very short, rudimentary, often reticulate down to the base, whitish, faintly pubescent, leathery then rigid and fragile. No special odor or flavor. Spores white, elliptical-elongated, smooth, 7–12 × 3–4 microns.

Edibility Usually too tough; of no interest.

Habitat Saprophytic and, more rarely, parasitic on various broadleaf species.

Season Most commonly seen in the spring.

Note The honeycomb-shaped pores are the most distinctive feature of this mushroom. Sometimes it lacks any stipe development at all, and what stipe there is is mostly covered with pores.

266 COLTRICIA PERENNIS

Synonym *Polyporus perennis.*

Etymology From Latin, "long-lasting," because the carpophores may live for one or more years.

Description Cap 2–8 cm, flattened or funnel-shaped, thin, finely velvety, then zoned and glabrous, rust-colored and brown. Tubes cinnamon-colored, slightly decurrent; pores quite broad, roundish or polygonal, initially whitish then cinnamon-brown, velvety. Stipe 1–3 × 0.5 cm, cylindrical, reddish brown, velvety. Flesh cinnamon-colored, leathery, thin. Odor mushroomy, flavor astringent. Spores from yellow to ochreous, elliptical, smooth, 5–10 × 3.5–6 microns.

Edibility Inedible because of consistency.

Habitat Frequent in burnt ground and places, in woodland, with caps often concrescent.

Season New growths summer to autumn; present all year.

Note *C. cinnamomea* is very similar but has a zoned, shiny, cinnamon-colored cap. Both species have cinnamon-colored to orange-brown flesh that blackens when touched with a drop of potassium hydroxide.

267   ONNIA TOMENTOSA

Synonym   *Polyporus tomentosus.*
Etymology   From Latin, "downy" or "velvety."
Description   Cap 5–10 cm, convex, flat or umbilicate, rust-brown, rust-red, cinnamon-yellow, not radially zoned, rugose, velvety, lighter at margin which is entire or lobate, fading to yellowish in age, sometimes with concrescent or fairly imbricate caps. Tubes decurrent, reddish brown; pores angular, irregular, often mazelike, tube-colored or more often grayish. Stipe 1–3 × 0.7–1.5 cm, central or otherwise, irregular, solid, spongy, soft, feltlike, downy, dark brown or blackish, sometimes almost absent. Flesh in two ochreous brown layers, the upper layer spongy, the lower firm, fibrous, and shiny when cut. Spores pale yellow in mass, elliptical, smooth, 4–5 × 3–3.5 microns.
Edibility   Too leathery, therefore inedible.
Habitat   On the ground or on roots of conifers, often seeming terrestrial but in fact growing from buried wood.
Season   Summer and autumn.
Note   An otherwise identical fungus, *Onnia circinata,* lacks a stipe and grows directly out of wood; it also differs somewhat microscopically.

---

268   PSEUDOHYDNUM GELATINOSUM

Synonym   *Tremellodon gelatinosum.*
Etymology   From Latin, "like gelatin."
Description   Cap 3–6 cm, fan-shaped, sessile or with a short stipe, surface papillose, brown to white. Hymenium with white translucent spines, conical, straight, 2–4 mm long. Stipe, when present, is short, squat, and whitish. Flesh transparent, thick, gelatinous. Odor insignificant, flavor distinctive but hard to describe. Spores white, globose, smooth, 14–18 × 10–12 microns.
Edibility   Good, especially when candied in sugar water, dried and dusted with confectioner's sugar.
Habitat   On coniferous stumps, in cold temperate zones.
Season   Summer to late autumn, winter.
Note   This fungus is easy to recognize because of its distinctive aculeate hymenium. It is a member of the jelly fungi, which have basidia and form fruit-bodies of varied shapes and colors.

269 AURISCALPIUM VULGARE

Etymology From Latin, "common."
Description Cap 1–2 cm, kidney-shaped to almost circular, convex, with the stipe attached laterally, villose, dark brown, sometimes zoned, edge ciliate, initially lighter. Fairly sparse aculei, at first violet-brown then whitish from the spores. Stipe 3–10 × 0.1–0.2 cm, cylindrical, with grayish hairs on a dark brown background. The base extends into a mycelial thread, surrounded by brown filaments. Flesh leathery, white, delimited by a black line from the hairy layer. Odor insignificant, flavor slightly acrid. Spores white, ovoid, finely warty, 4–6 × 3–4 microns.
Edibility Of no interest because of size.
Habitat On pinecones, widely distributed in the Northern Hemisphere but not usually common.
Season All year, but particularly after the first spring rains.
Note Very often grows out of moss, beneath which there is a host cone.

---

270 HYDNUM IMBRICATUM

Synonym *Sarcodon imbricatum.*
Etymology From Latin, "covered with tiles," because of the appearance of the cap.
Description Cap 6–30 cm or more, convex then flat, often slightly umbilicate, eventually funnel-shaped, floccose, tessellated and squamose with large gray-brown scales, persistent or slightly caducous. Teeth decurrent, ash-white then brown, 1–1.2 cm long. Stipe 2.5–7.5 × 2.5–5 cm, short, thick, smooth, whitish or cap-colored. Flesh whitish then light gray-brown, thick, consistent, sometimes zoned. Odor slightly iodized, sometimes horselike, flavor astringent or slightly bitter. Spores reddish brown, oval or subglobose, warty, 5–6 × 5 microns.
Edibility Fair to good, best when reduced to powder and used for flavoring.
Habitat In conifer and mixed broadleaf-conifer woods; often in groups.
Season Spring to autumn.
Note *H. scabrosum* is intensely bitter, has a less coarsely scaly cap, and a stipe that becomes dark greenish black at the base.

271 HYDNELLUM SCROBICULATUM

Synonym *Hydnum ferrugineum.*
Description Cap 3–8 cm, first hemispherical-convex, soon expanding, irregular, humped, markedly downy and feltlike, roughening with protruding knobs and bumps, first white with drops of red exudate, then brownish yellow or brownish red with an oblong margin, whitish then cap-colored. Teeth up to 0.5 cm long, unequal, decurrent, crowded, white then pinkish with brown spots, finally entirely rust-brown, fragile. Stipe 2–3 × 1–2 cm, irregular, cap-colored, downy, enlarged at spongy base. Flesh spongy then suberose, pinkish brown in cap with faint concentric zonation, pale purple-brown in stipe. Faint mealy odor, flavor sweetish. Spores light brown, subglobose or broadly elliptical, sometimes polyhedral, 4.8–5.6 × 3.5–4.5 microns.
Edibility Inedible because of consistency.
Habitat Gregarious, often concrescent, on the litter in coniferous woods, especially spruce.
Season Summer and autumn.
Note The flesh reacts violet then olive-green with potash (potassium hydroxide). This species occurs in eastern and central North America.

272 DENTINUM REPANDUM

Synonym *Hydnum repandum.*
Etymology From Latin, "bent backwards."
Description Cap 5–15 cm, convex then open, irregular, often depressed, margin incurved, lobate, undulate, surface dry, adherent, typically orange but sometimes pale yellow to whitish. Teeth up to 6 mm long, fragile and easily detachable, cream-colored or reddish ochre. Stipe 3–8 × 0.5–3 cm, fairly slender, solid, quite irregular, flared at top, whitish, sometimes with brown markings, downy at base. Flesh thick, consistent, white turning pinkish yellow, fragile. Odor slightly fruity, flavor sweet to slightly bitter. Spores white, subglobose, smooth, 6–7 × 5–6 microns.
Edibility Fair to very good.
Habitat Gregarious, sometimes concrescent, beneath broadleaf trees and conifers.
Season Spring and late autumn.
Note In parts of northeastern North America this is considered an excellent edible; elsewhere it seems not to be used much at all. There are distinct color varieties, some of which may be better than others, and there is a similar species, *D. umbilicatum*, which is smaller and umbilicate, and has larger spores.

273   LEOTIA LUBRICA

Etymology   From Latin, "viscid" or "slippery."
Description   Carpophore 5–20 mm wide, irregularly rounded, fairly lobate, separated from stipe by a fairly deep vallecule, greenish orange, darker with age, smooth, slightly viscous in wet weather. Stipe 1–5 × 0.2–0.5 cm, cylindrical or slightly enlarged at the base, hollow when mature, ochreous yellow or orange tending to become greenish, viscous. Flesh soft, almost gelatinous in the head, waxy in the stipe. No particular odor or flavor. Spores colorless under microscope, elliptical, slightly curved, smooth, 20–25 × 5–7 microns.
Edibility   Of no value because of size and texture.
Habitat   In small groups in woods, on the ground, amid ferns or moss.
Season   Summer and autumn.
Note   *L. viscosa* has a dark green head and buff- to orange-colored stipe; *L. atrovirens* has both a green head and stipe, although often of different hues.

274   HELVELLA ELASTICA

Synonym   *Leptopodia elastica.*
Etymology   From Greek, "pushing" or "elastic," referring to the texture of the flesh.
Description   Carpophore 4–8 cm high, two-lobed head, undulate, free, at most 4 cm wide, whitish turning brown with age and dry weather, grayish ochre when mature. Stipe 2.5–7.5 cm × 2–10 mm, slightly supple, pruinous when young, white, sometimes with reddish marking at base when mature, fairly compressed. Flesh whitish, elastic. No particular odor or flavor. Spores whitish, elliptical, smooth, 21–27 × 13–16 microns.
Edibility   Fair when cooked.
Habitat   In cool shady parts of woods, among leaves.
Season   Spring to autumn.
Note   This is one of the more common species in the genus. Many are small and most are difficult to identify without a microscope. Although many are well known and eaten in both Europe and North America, **none should ever be eaten raw.**
Caution   Be very careful with this group because of their close relationship to some seriously poisonous species of *Gyromitra* or false morels.

275 HELVELLA LACUNOSA

Etymology From Latin, "with holes," because of the structure of the stipe.

Description Carpophore 5–12 cm high, head 2–4 cm wide, normally three-lobed, blackish, fairly undulate, margins sometimes partly joined at stipe. Stipe 4–10 × 1–2 cm, dark gray to blackish, deeply sulcate lengthwise, fistular at the center of the flesh. Flesh quite thin and elastic. Pleasant mushroomy odor and flavor. Spores whitish, elliptical, smooth, 18–20 × 12–13 microns.

Edibility Good when cooked.

Habitat On fine, sandy ground, especially in open woodland near paths.

Season Spring to autumn, over winter in California.

Note Although large and common in some areas, in others it is rarely found and easily overlooked. As with other species of *Helvella*, **great caution should be exercised** identifying or sampling this mushroom.

---

276 HELVELLA MONACHELLA

Etymology From Greek, "small nun," from its resemblance to the habit of certain Catholic religious orders.

Description Carpophore not wider than 4 cm, head with two or, more frequently, three lobes, partly joined together to form a typical mitre shape, brown, blackish brown uppermost, faintly brownish or finely pubescent below. Stipe 4–12 × 0.8–2 cm, slightly club-shaped, whitish or later faintly ochreous, from pubescent to glabrous, fistular, enlarged at the base. Flesh whitish, slightly elastic. No special odor or flavor. Spores whitish, elliptical, smooth, 21–22 × 14–15 microns.

Edibility Quite good when cooked.

Habitat Gregarious on sandy ground, preferably beneath poplars.

Season Spring.

Note This dark species of *Helvella* is not known to occur in North America.

277 HELVELLA CRISPA

Etymology From Latin, "curled," from its shape.

Description Carpophore 5–15 cm high, head 3–5 cm wide, kidney- or saddle-shaped or lobate, fairly undulate and curled, edge free, white or whitish, glabrous, ochreous and pale beneath, venose and villose. Stipe 5–12.5 × 2–3 cm, deeply sulcate lengthwise, lacunose, whitish, barely pubescent when young, then glabrous. Flesh whitish, quite elastic. Mushroomy odor and flavor. Spores whitish, elliptical, smooth, 20–22 × 12–13 microns.

Edibility Good when cooked, although somewhat chewy.

Habitat On fine ground by paths and in open woodland.

Season Spring, but more frequent in autumn.

Note This somewhat resembles *H. lacunosa* (**275**) except for its color, and that its head is more saddle-shaped than lobed.

Caution is recommended because some saddle-shaped mushrooms are believed to be poisonous, and are closely related to the dangerous false morels, species of *Gyromitra*.

---

278 GYROMITRA INFULA

Synonym *Helvella infula.*

Etymology From Latin term describing a turbanlike headdress worn by Roman priests and vestal virgins.

Description Carpophore with saddle-shaped head with 2, 3, or 4 lobes, 6–10 × 8–12 cm, enlarged, rugose, irregular, ending in a point, external surface fertile, plicate, pinkish brown in wet weather, brownish yellow when dry, darker and brown with age. Stipe 6–10 × 1.5–3 cm, cylindrical then compressed and deeply sulcate, hollow, pruinous, pale brown then reddish pink or lilac-gray, yellow at the base. Flesh soft, waxlike. Faint mushroomy odor and flavor. Spores pale yellow, elliptical, smooth, 19–24 × 7–8 microns.

Edibility Eaten in Europe, but not recommended in North America because not enough is known about this species and its look-alikes.

Habitat In pine and beechwoods, among remains of wood, in charcoal kilns or burnt areas, also on sawdust.

Season Summer and autumn.

Note Although saddle-shaped, this is placed with the brain-shaped mushrooms in the genus *Gyromitra*. *G. ambigua* is shorter, its stipe is tinged with violet, and it is more common in colder regions.

Caution No *Gyromitra* should be eaten in any form.

279 GYROMITRA ESCULENTA

Synonym *Helvella esculenta.*
Etymology From Latin, "edible."
Description Carpophore up to 10–12 cm high, head 5–10 cm, subglobose, irregularly lobate and circumvolute, brownish. Stipe 2–5 × 2–4 cm, compressed or irregularly plicate, whitish or tinged with pink or violet, glabrous or pruinous the whole length. Flesh whitish, waxy, fragile. Strong mushroomy odor, flavor sweetish. Spores whitish, elliptical, smooth, 17–21 × 9–11 microns.

Edibility Rated by some a good edible in Europe and North America, this fungus can cause life-threatening illness if toxins are not thoroughly removed by cooking. Some ways of preparation will reduce or eliminate the chances of an acute poisoning, but this mushroom is also known to contain a potent carcinogen.
Habitat Typically in mountainous pinewoods.
Season Spring and early summer.
Note Acute poisoning from this mushroom usually occurs about six hours after ingestion, and symptoms include a bloated feeling, cramps, vomiting, diarrhea, convulsions, coma, and death.

280 VERPA CONICA

Synonym *Verpa digitaliformis.*
Etymology From Latin, "conical."
Description Head 2–3 × 1.5–2.5 cm, campanulate, obtuse and slightly depressed at top, slightly sulcate, brownish on fertile surface, whitish or yellowish in sterile part, margin white, fairly adherent to stipe, but always free, hollow. Stipe 5–7 × 0.8–1 cm, cylindrical, slightly narrowing at top, almost hollow, whitish or tinged with pink-ochre, with small, slightly darker, adpressed scales. Flesh white, watery, waxy, fragile. Odor and flavor slight. Spores white, elliptical, smooth, 23–30 × 13–17 microns.

Edibility Fair.
Habitat At edge of woods or in hedges, especially under the Rosaceae (rose family).
Season Early spring.
Note This is one of the first ground mushrooms to appear in the early spring. It is rarely found in quantity, and is easily overlooked; it appears before most mushroom hunters are out looking for morels.

281 VERPA BOHEMICA

Synonym *Ptychoverpa bohemica.*
Etymology From Latin, "Bohemia."
Description Head 2–5 × 3–4 cm, campanulate or conical-ovoid, free, yellowish brown, covered with lengthwise ribs, ramified, undulate in specimens growing in wet conditions, straight when dried, forming irregular alveoli, elongated or straight, hollow. Stipe 6–8 × 0.8–1 cm, coarsely cylindrical, narrowing toward top, smooth, white, fistular, solid. Flesh white, waxy, watery, fragile. Odor slightly acidic, flavor slightly bitter but not unpleasant. Spores white, elliptical, sometimes slightly curved, smooth, 60–80 × 18–20 microns.
Edibility Fair, though slightly watery. Caution is advised, however, because immoderate ingestion of this species can cause some temporary but uncomfortable muscular incoordination.
Habitat In the leaf litter beneath broadleaf trees, especially poplars.
Season In very wet spring periods.
Note This mushroom can be differentiated from morels by examining the attachment of the head to the stipe: in morels, the attachment is at the base of the head; in *Verpa*, it is at the apex of the head.

282 MORCHELLA SEMILIBERA CRASSIPES

Synonym *Mitrophora crassipes.*
Etymology From Latin, "swollen-footed."
Description Carpophore up to 20 cm high, head conical, dark olive-brown, pointed at top, with long primary ribbing, angular, and a few irregular secondary alveoli, margin free, quite undulate. Stipe 5–12.5 × 1–2 cm, with fairly enlarged base, sulcate, hollow, attached midway up head, surface at first white, markedly furfuraceous-squamose, then slightly orange-ochre. Flesh thin, whitish. No particular odor or flavor. Spores whitish to cream-colored, elliptical, smooth, 22–30 × 14–18 microns.
Edibility Good.
Habitat Gregarious, on fine humus-rich ground in rows of trees and hedges with elm.
Season Spring.
Note This is a variety or only a variant form of *M. semilibera* (**283**). The size and shape of morels are affected by habitat, rainfall, and whether they appear at the beginning or end of their season: later morels are often larger and thicker. *M. crassipes*, which is thick-footed, differs from this morel by the attachment of its stipe to the margin of the head.

283 MORCHELLA SEMILIBERA

Synonyms *Morchella hybrida; Mitrophora semilibera.*
Etymology From Latin, "half-free," because the lower part of the head does not touch stipe.
Description Carpophore 5–15 cm high, head conical, pointed or obtuse at top, free in lower half, olive-brown, ochreous in dry weather, with lengthwise ribbing, sinuate then symmetrical, often anastomosed, with secondary four-sided alveoli, hollow. Stipe 5–12.5 × 1–2 cm, white or light ochre, slightly enlarged at base, furfuraceous, hollow. Flesh thin, firm. No particular odor or flavor. Spores whitish, elliptical, smooth, 22–30 × 13–17 microns.
Edibility Good.
Habitat Single to several in leaf litter under broadleaf trees.
Season Spring.
Note This fungus is distinguished from other morels by the free margin of its head; it is attached to the stipe about midway up the head. It usually follows the appearance of the *Verpas* and precedes and overlaps the appearance of the various true morels.

---

284 MORCHELLA ELATA

Etymology From Latin, "slender."
Description Carpophore 6–12 cm high, head conical or cylindrical, blackish, divided by lengthwise blackish ribbing connected by blackish ribs, thus forming fairly large alveoli, fairly rectangular, hollow. Stipe 5–7.5 × 1–2 cm, almost cylindrical, ochreous gray, sometimes sulcate, conspicuously furfuraceous, hollow. Flesh quite thin and tough, whitish or grayish. No particular odor or flavor. Spores white, elliptical, smooth, 25–27 × 16–18 microns.
Edibility Good.
Habitat Beneath poplars and pines.
Season Spring.
Note This is a blackish mushroom easily recognized by its radially arranged longitudinal ribbing. Because of unusual variation in color, shape, and size, morels are difficult to identify to species. But all species of *Morchella*, true morels, are edible and very good when cooked and eaten in moderate amounts. The photograph represents *Morchella elata* var. *purpurascens*, a European variant.

285 MORCHELLA CONICA

Etymology From Latin, "cone-shaped."
Description Carpophore 5–10 cm high, head pointed, conical, olive-ochre to grayish black, varying in color with age and moisture, with elongated alveoli and lengthwise ribs often as long as the head, which soon become blackish; crosswise ribs divide the primary alveoli into secondary, almost quadrangular alveoli, hollow. Stipe 5–7.5 × 1–2 cm, pale yellowish, furfuraceous, almost cylindrical, hollow. Flesh quite thin and tough, whitish. No particular odor or flavor. Spores whitish, elliptical, smooth, 22–24 × 12–15 microns.
Edibility Good.
Habitat On acid or burnt ground, beneath conifers.
Season Spring.
Note Like all the species of *Morchella*, in addition to the typical form there are a great many varieties, each with distinctive color and appearance. The specimen in the photograph belongs to the variety *distans*.

---

286 MORCHELLA ESCULENTA

Synonym *Morchella rotunda*.
Common name Morel.
Etymology From Latin, "edible."
Description .Carpophore 10–20 cm high or more, head normally rounded, sometimes slightly conical, ochreous yellow, with large alveoli, irregularly rounded, slightly venose, separated by sterile paler ribs, hollow. Stipe 5–15 × 1–2 cm, sometimes up to 5 cm at base, strong, hollow, pale, barely furfuraceous, enlarged and sulcate at base. Flesh thick, tender, whitish. Strong mushroomy odor, sweetish flavor. Spores white, elliptical, smooth, 20–23 × 12–13 microns.
Edibility Excellent; also good for drying.
Habitat On sandy or clayey-sandy ground, in open places, especially near tulip-poplar, ash, and dead elm, and in old apple orchards.
Season Spring.
Note Morels are especially variable in shape, size, and color, and many species described in the literature, such as *M. deliciosa*, may prove to be only variants of this one very common polymorphic species.

287　MORCHELLA UMBRINA

Etymology　From Latin, ''dark.''
Description　Carpophore 3–7 cm high, head globose, up to 3 cm in diameter, blackish gray, initially with primary adpressed ribs, fairly sinuate, then dilated, with numerous deep alveoli, first roundish then irregular. Stipe 2–2.5 × 1 cm at top, conspicuously enlarged at base, where it is also sulcate and cavernous, almost glabrous, whitish, sometimes with ochreous markings, hollow like the fertile part. Flesh quite thick, white to brownish. No particular odor or flavor. Spores whitish, elliptical, smooth, 18–23 × 9–12 microns.
Edibility　Excellent.
Habitat　On fine sandy ground, in sunny places, near poplars or other broadleaf species.
Season　Spring.
Note　Unfortunately, this delicious morel is not known to occur in North America. The clear contrast between the blackish head and the white stipe make this one of the most easily identifiable morels. The fungi belonging to the genus *Morchella* have a wide variety of carpophores, which often look as if they belong to more than one species.

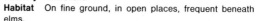

288　MORCHELLA VULGARIS

Etymology　From Latin, ''common.''
Description　Carpophore up to 10–15 cm high, head elongated, rarely roundish, blackish gray, with elongated alveoli, quite irregular, with thick lengthwise ribbing, sometimes with reddish markings, hollow. Stipe 7.5–10 × 1–2 cm, whitish, almost glabrous, enlarged and sulcate at the base, hollow. Flesh tender, whitish. Mushroomy odor and sweetish flavor. Spores whitish, elliptical, smooth, 18–20 × 10–12 microns.
Edibility　Excellent.
Habitat　On fine ground, in open places, frequent beneath elms.
Season　Spring.
Note　This mushroom is recognized in Quebec but not elsewhere in North America. It differs from *M. esculenta* (**286**) by its darker color and more elongated shape, and from such black morels as *M. conica* (**285**) and *M. elata* (**284**) by the flush attachment of its head to the stipe; the others have a slight overhang or depression between the margin of the head and the stipe.

SHELF OR CRUST
MUSHROOMS

289 PANUS CONCHATUS

Synonym *Panus torulosus.*
Etymology From Latin, "shell-shaped."
Description Cap 5–10 cm, fairly funnel-shaped, eccentric, or fan-shaped, irregularly undulate, first entirely amethyst-violet, or only at edge, with center then completely ochreous yellow, first glabrous, with radial fibrils, innate or finely pubescent, downy at margin, squamulose when mature. Gills whitish, finally ochreous, tinged with amethyst on or toward edge, very decurrent in parallel rows, with some ramifications, edge curling as it dries. Stipe 1–3 × 0.5–1 cm, eccentric or lateral, light ochre, irregular, compressed with downy base. Flesh white, limp but leathery, especially when mature. Sometimes a slight aniseedlike odor and faint earthy flavor. Spores white, elliptical, smooth, 4–7.5 × 2–3.5 microns.
Edibility Can be eaten but of poor quality.
Habitat Preferably beneath broadleaf species, mainly beech, poplar, and willow.
Season Late spring, summer, and autumn.
Note Widely distributed but not very common. Much more common is *P. rudis*, which is pinkish to reddish brown, fading to ochre, tannish or white on drying, and densely covered with long stiff hairs.

290 CREPIDOTUS MOLLIS

Synonym *Crepidotus fulvotomentosus.*
Etymology From Latin, "soft."
Description Cap 3–7 cm, almost sessile, bracket- or kidney-shaped with a tiny lateral stipe, glabrous and covered with sparse brown or yellowish hair on a pale olive-gray background, then whitish, pale yellow, undulate and lobate when mature. Gills whitish gray then light brown, convergent and decurrent at base, often ramified. Stipe, when there is one, up to 1.5 cm long, villose. Flesh whitish, soft, almost gelatinous beneath cuticle. No odor, flavor sweet. Spores ochreous brown, elliptical, smooth, 7–9 × 5–6 microns.
Edibility Fair.
Habitat In groups, often imbricate, or isolated on trunks, fallen branches, and sawdust.
Season From spring to the first winter frosts.
Note *Crepidotus* is a genus of mostly stipeless brown-spored mushrooms that resemble small, fragile species of *Pleurotus*. There are a great many species in North America; most are small, very few are common. Because their identification to species is difficult without a microscope, and because little is known about the edibility of the various species of *Crepidotus*, none should be casually gathered for the table.

291 MERULIUS TREMELLOSUS

Etymology From Latin, "trembling," because of its gelatinous appearance.

Description Carpophore crustaceous, or very adpressed to the substratum, margin free, bracket-shaped, connate and also imbricate, sterile surface free and downy, whitish, with a pinkish, orange, sinuate, denticulate margin. Hymenium from pinkish white to bright orange, with alveolar folds 1–3 mm wide, or linear, supple, anastomosed. Flesh gelatinous, elastic-leathery, cartilaginous when dry. Spores white, cylindrical-arcuate, smooth, 3.5–4.5 × 1–1.2 microns.

Edibility Too leathery to eat.

Habitat On rotted broadleaf and coniferous wood.

Season All year, especially in autumn.

Note This is the most common and most widely distributed species of *Merulius*, a genus of crust fungi that are sometimes shelving and fleshy, and that resemble a cross between the jelly fungi and polypores.

---

292 SCHIZOPHYLLUM COMMUNE

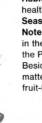

Etymology From Latin, "common," or "widespread."

Description Cap 1–3 cm, gray or flesh-colored, tending to become white, fan- or kidney-shaped, fairly lobate, sessile, dry, inclined downward or horizontal, covered with evident grayish white hairs, sometimes zoned concentrically. Gills grayish white or pink-tinged, split lengthwise into two layers which tend to curve away from each other, converging radially toward base, where they are attached. Flesh brownish, becoming whitish and sometimes malodorous as it dries. Spores white, straight or curved, smooth, 3–4 × 1–1.5 microns.

Edibility Of no interest because of texture.

Habitat On wood, stumps, dead trunks or living trees in poor health.

Season All year.

Note This is probably the most widely distributed mushroom in the world; there seems to be no place except in water or at the Poles where it is not found, and it is usually very common. Besides growing on wood, it has also been found on animal matter (saprophytic on whalebone or, in mycelial form with no fruit-body, on scar tissue in a human mouth).

293 DAEDALEA QUERCINA

Etymology From Latin, "pertaining to oak," because of its habitat.

Description Carpophore bracket-shaped, 9–50 cm, sessile, very rarely crustaceous or with a hint of a stipe, light brown, smooth or rough, quite rugose, unequal, marked by alternate concentric zones, raised and depressed. Pores 6–50 mm long, sinuate, gill-like, ramified and anastomosed, thick, ligneous, brown-gray or leather-colored. Flesh pale reddish brown or light brown, ligneous, thick, quite light. Mushroomy odor. Spores white, ovoid, smooth, 6 × 2–3 microns.

Edibility Too leathery to eat.

Habitat Saprophytic, preferably on oak.

Season All year.

Note A perennial, worldwide species, varying somewhat in the formation of the hymenium and in the presentation of the carpophore. The mazelike hymenial structure is also found in *Daedaleopsis confragosa*, which has thinner pore walls and smaller pores that bruise pinkish on handling.

294 LENZITES BETULINA

Etymology From Latin, "of birch," because of its preferred habitat.

Description Carpophore 2.5–10 cm, bracket-shaped, usually flat, sessile, sometimes crustaceous, upper surface downy, with not very conspicuous zonation, but sometimes with concentric zones, basic color grayish white, with a tendency to turn paler and become uniform. Hymenium with dirty white gill-like plates which part at the base, straight and single, or ramified, often anastomosed, sharp-edged. Flesh white, suberose, leathery. Spores white, cylindrical, smooth, 4–7 × 2–3 microns.

Edibility Inedible because of the leathery texture.

Habitat Frequently on birch, but also on a wide variety of dead broadleaf wood, even construction timbers and telegraph poles.

Season All year.

Note Similar species include *Gloeophyllum sepiarium* (**295**) and *G. abietinum,* which grows on firs and other conifers, and is brownish with reddish gray-brown, pruinose, gill-like plates.

295 GLOEOPHYLLUM SEPIARIUM

Synonym *Lenzites sepiaria.*
Etymology From Latin, "of fences," because of its habitat.
Description Carpophores bracket-shaped, very broad, 2–10 cm, sessile, isolated or superposed or joined laterally to each other, quite often at the back, thin, velvety, with radial wrinkles and concentric grooves, brownish yellow then dark red, finally blackish brown, margin regular or slightly undulate, downy, orange-yellow. Hymenium mazelike with gill-like plates radially arranged, crowded, anastomosed, ligneous, not separable, brownish ochre, edge at first pruinous, whitish then reddish brown. Flesh thin, suberose, tobacco-colored. No particular odor or flavor. Spores white, smooth, cylindrical, 7.5–8.5 × 2.5–3 microns.
Edibility Inedible because of texture.
Habitat On coniferous wood in open, sunny places.
Season All.year.
Note It often attacks wood put to human use, such as telegraph poles, fences, etc., causing the wood to rot and turn red.

---

296 PHAEOLUS SCHWEINITZII

Synonym *Polyporus schweinitzii.*
Etymology After the mycologist Schweinitz.
Description Carpophore 10–40 cm, typically bracket-shaped and imbricate, initially leather-colored, then darkening toward the edge, rust-brown in central part, remaining yellow at margin, surface rugose, feltlike, downy. Tubes decurrent, yellow to rust-brown; pores irregular, polygonal, sometimes mazelike. Stipe, when present, 3–12 × 5–6 cm, rust-brown, solid, rugose, sometimes covered with tubes to the base. Flesh rhubarb-colored, tending to darken, spongy then fibrillose, finally dry and fragile. Spores white, elliptical, smooth, 7–8 × 4 microns.
Edibility Inedible because of texture.
Habitat On coniferous stumps and roots.
Season Summer and autumn.
Note This fungus is commonly found with grasses growing up through the fruit-body.

297 PIPTOPORUS BETULINUS

Synonym *Polyporus betulinus.*
Etymology From the Latin, "of birch," because of its habitat.
Description Carpophore 4–20 cm or more, up to 6 cm thick, rounded or kidney-shaped, sometimes with a short peduncle, pale tending to darken with age, often dappled, covered with a thin, smooth, separable pellicle, margin rounded, obtuse, and sterile. Tubes white, 2–8 mm, sometimes detachable; pores white, darkening, small and round. Flesh white, soft, suberose, crumbly when old. No odor, flavor bitterish. Spores white, cylindrical, often curved, smooth, 5–7 × 2 microns.
Edibility Mediocre, can only be eaten when very young as it is soon too tough and unpalatable.
Habitat Parasitic on birch; rarely, on beech.
Season Grows in summer and autumn, but present all year.
Note This is a very common, conspicuous mushroom over winter throughout the northern range of birch woodlands, sometimes covering the sides of recently dead, white birch trees.

---

298 PYCNOPORUS CINNABARINUS

Synonym *Polyporus cinnabarinus.*
Etymology From Greek, "like cinnabar" or "dragon's blood," because of its color.
Description Carpophore 3–6 cm or more, bracket-shaped, sessile, deep orange-red, tending to darken, at first slightly pubescent then glabrous, fairly rugose, with faint zonation toward margin. Tubes 1–3 mm long, blood-red; pores small, round, pubescent, vermilion. Flesh red, leathery, first spongy then suberose. Odor and flavor negligible. Spores white, cylindrical, smooth, 5–6 × 2–2.5 microns.
Edibility Inedible because of toughness.
Habitat On dead broadleaf branches and trunks.
Season Summer and autumn.
Note This is an easily indentifiable polypore, often common in late fall and readily seen at a distance because of its bright cinnabar-red color. In the Southeast, there is a more red, less orange species, *P. sanguineus,* which often looks shiny or satiny; it is also quite common throughout the littoral forests of the West Indies.

299 PHELLINUS PUNCTATUS

Etymology   From Latin, "spotted" or "dotted."
Description   Carpophores crustaceous, up to 20 cm long and 0.5–2.5 cm thick, stratified, cinnamon-colored or brownish; pores up to 0.7 cm long, fine, roundish, cinnamon- or tobacco-colored, with a grayish, hazel pruinescence, mycelium pale or sulfur-yellow. No particular odor or flavor. Spores white then pale yellow, oval, subglobose, smooth, 6.5–9 × 5–8 microns.
Edibility   Inedible because of texture.
Habitat   On all types of broadleaf wood or bushy shrubs, sometimes parasitic on vines.
Season   All year.
Note   *Phellinus* is a genus of corky polypores that are less woody than similar forms. The species can be crustlike or shelflike, the flesh is some shade of brown, and most species contain setae (microscopic, darkly pigmented, awl-shaped, sterile cells). Common species include *P. gilvus,* with mustard yellow flesh and dark brown overlapping or crustlike caps; and *P. igniarius,* a perennial bracket on broadleaf trees, with orange-brown flesh and tubes stuffed with an abundant, thready white mycelium.

300 TRAMETES VERSICOLOR

Synonyms   *Polyporus versicolor; Coriolus versicolor.*
Etymology   From Latin, "varying in color."
Description   Carpophores 3–8 cm wide, flattened or slightly depressed at the attachment, thin, in superposed brackets or joined together to form a roselike structure, sessile, smooth or velvety, with variously colored zonation, separated by satinlike shiny zones. Tubes short, whitish, darkening to brownish on maturity; pores small, round, initially white then light brown. Flesh thin, leathery, whitish. No odor or flavor. Spores white, cylindrical or slightly arcuate, smooth, 4.5–8 × 1.5–3 microns.
Edibility   Inedible because of texture.
Habitat   On dead and living trunks and wood, both coniferous and broadleaf.
Season   All year.
Note   This is a common species, quite variable in color, but with the upper surface invariably villose and conspicuously zoned with different colors.

301 FOMES FOMENTARIUS

Synonym *Ungulina fomentaria.*
Etymology From Latin, "producing tinder for fire," because of its use.
Description Carpophore 10–60 cm, bracket-shaped or hoof-shaped, sessile, grayish to gray-brown with faint semicircular markings, the most conspicuous of which is in a state of growth; margin cream-colored, hazel, or light brown. Tubes 1–3 cm long, rust-colored, many-layered; pores small, round, pruinous, gray or light hazel. Flesh brown, suberose, soft, 7–20 cm thick, with annual series of stratified tubes, beneath a hard, thick crust, shiny blackish gray in cross section. During its growth it develops a marked bananalike odor. Spores white, elliptical, smooth, 14–22 × 5–7 microns.
Edibility Inedible because too woody.
Habitat Parasitic on various broadleaf trees.
Season This species lives for several years, is present all year, and grows from spring to autumn.
Note The flesh of the pulverized carpophore was used in ancient surgery as a styptic agent; and when mixed with saltpeter it produced the best tinder for lighting fires.

302 FOMITOPSIS PINICOLA

Synonym *Ungulina marginata.*
Etymology From Latin, "pine."
Description Carpophore 10–30 cm, hoof-shaped, sessile, thick or flattened, yellowish then red, finally black, with small semicircular concentric markings, covered in the older parts by a resinous, blackish crust, shiny or pruinous, margin yellow or red. Tubes ochreous, stratified, 3–6 mm long; pores round, small, cream or very light brown, sometimes reddish when rubbed. Flesh white or pale yellow, suberose, hard with stratified tubes. Odor of tobacco, flavor slightly acid and bitter. Spores pale yellow-white, elliptical, smooth, 6–10 × 3–4.5 microns.
Edibility Inedible because too woody.
Habitat Parasitic on broadleaf and coniferous species, also saprophytic on dead tree trunks.
Season Perennial, growing from spring to first cold spell in autumn.
Note This species varies in accordance with the period of growth and also, apparently, on the basis of the host. Easy to identify because of the bright red obtuse margin, which turns from red to orange-yellow at the edge and pale yellow lower down. Despite its name, it has a large range of host trees.

303   GANODERMA APPLANATUM

Common name   Artist's conk.
Etymology   From Latin, "flattened."
Description   Carpophore 10–40 cm, flat bracket-shaped, rarely hoof-shaped, often imbricate, sessile, sometimes first white, soon covered by a smooth, yellowish crust, becoming reddish brown, appears varnished, smooth or powdery, marginal area white or grayish. Tubes rust-colored, stratified, 1–4 cm long; pores white, turning brown when touched, small, round or irregular. Flesh brown, fairly dark, with whitish zones, feltlike. Spores rust-brown, elliptical, smooth or slightly and finely warty, 9–13 × 6–8 microns.
Edibility   Inedible because too woody.
Habitat   Parasitic, persisting after death of the host as a saprophyte, especially on broadleaf trees.
Season   All year, perennial.
Note   This fungus is easily identified by its flat shelves, somewhat crusty surface, and white, readily discoloring pore surface. It is often called artist's conk because a picture can be etched on the whitish pore surface; wherever touched it turns brown and, if left to dry, will retain a drawing.

304   GANODERMA TSUGAE

Description   Carpophore 4–20 cm, fan- or kidney-shaped, surface glabrous, unequal, concentrically sulcate, appears varnished, vermilion yellow-red to red-black or blackish with margin acute, bright yellow, finally cap-colored and radially sulcate, often undulate or fairly lobate. Tubes 0.5–0.7 cm long, brown, not detachable; pores circular or polygonal, from white to light cinnamon-colored. Stipe 2–20 × 1–4 cm, sometimes very small, lateral, cylindrical, often forked, similar in color and surface formation to cap. Flesh suberose, almost ligneous, fibrous, white or whitish beneath bark, hornlike appearance. Spores ochreous brown, ovoid, warty, 9–11 × 6–8 microns.
Edibility   Edible and tender when it first forms in the spring, but soon becomes encrusted.
Habitat   On trunks and stumps of eastern hemlock (*Tsuga canadensis*) and other conifers.
Season   An annual species, which grows in summer and autumn, fairly persistent.
Note   The stipe may be fairly well developed and may be either on the same level as the cap or perpendicular to it and erect. Other species include *G. lucidum,* common on broadleaf trees, especially maple, and *G. oregonense,* a Northwest species that is larger and has a darker, brownish, less shiny cap.

305 HETEROBASIDION ANNOSUM

Synonym *Formes annosus.*
Etymology From Latin, "with many years."
Description Carpophore 7–45 cm, bracket-shaped, sessile, brown then blackish, convex or almost flat, sometimes barely imbricate, often resupinate or with a false stipe; surface uneven, rugose, sulcate, becoming crustaceous, smooth, margin thin, first white. Tubes yellowish, stratified; pores roundish, slightly angular, white then whitish or yellowish. Flesh whitish, suberose, then ligneous. No particular odor or flavor. Spores white, subglobose, minutely spiny, 4.5–6 × 3.5–4.5 microns.
Edibility Inedible because too woody.
Habitat Parasitic on conifers, rarely on broadleaf species.
Season A perennial species which grows in summer and early autumn.
Note The mycelium parasitizes the root apparatus of many conifers, causing serious damage. The fruit-bodies may also form on the ground in association with roots, or at the foot of trunks and stumps. The trees attacked by this fungus are easily blown down by wind.

306 INONOTUS HISPIDUS

Synonym *Polyporus hispidus.*
Etymology From Latin, "hirsute" or "hairy."
Description Carpophore 10–30 cm or more, bracket-shaped, sessile, very bristly, rust-yellow turning brown and finally blackish with age. Tubes 2–3 cm, rust-colored; pores small, round, brown, initially with a yellowish edge and small brown droplets of exudate. Flesh up to 10 cm thick, rust-colored, spongy, fibrous, becoming fragile when dry. Aromatic mushroom odor when young, flavor quite astringent. Spores ochreous brown, subglobose, smooth, 9–10 × 7–8 microns.
Edibility Inedible because astringent and too fibrous.
Habitat Parasitic on various broadleaf species, including ash, mulberry, oak, and willow; also reported on pine and fir. It occurs as far north as Massachusetts, and throughout the Southeast, west to Texas, New Mexico, and California.
Season Summer and autumn, but present all year.
Note *I. cuticularis* is similar but smaller, has a fibrillose-to-mentose rather than bristly surface, and smaller, elliptical spores.

307 FISTULINA HEPATICA

Common names   Ox-tongue fungus; poor man's beefsteak; beefsteak fungus.

Etymology   From Greek, "liverlike."

Description   Carpophore 5–30 cm or more, bracket-shaped, spatula-shaped or roundish, sessile or with a short lateral stipe, blood-red or liver-red, surface thick and gelatinous, quite viscous; as it ages it becomes blackish-brown and dries out. Tubes fine, pale, reddening, separate; pores round, cream-colored then reddish. Flesh reddish, marbled with pale purple lines, thick, fibrous, succulent. Slightly acid odor and flavor. Spores pink, subglobose, smooth, 4.5–6 × 3–4 microns.

Edibility   Good, cooked and raw.

Habitat   On broadleaf trunks and stumps, especially oak and chestnut.

Season   Summer and early autumn.

Note   Unlike most polypores whose tubes are fused, this fungus has clearly separated tubes. Since the loss of the American chestnut, it is now found mainly on oak stumps or at the base of diseased trees. It is mostly confined to eastern North America and, while usually not common, can be abundant. It is a good edible, especially raw: it has a lemony tart flavor and pleasing texture.

308 TYROMYCES CAESIUS

Synonym   *Polyporus caesius.*

Etymology   From Latin, "blue-gray."

Description   Carpophore 1–8 cm, bracket-shaped, often imbricate, sometimes with a short stipe, or crustaceous; white then blue-gray, surface villose, silky in old specimens. Tubes 3–9 mm long, white, not stratified; pores white, turning blue when touched, small, unequal, mazelike or aculeiform. Flesh white, blue when broken, soft, watery then brittle. Sweet odor and flavor. Spores light blue, elongated, often curved, smooth, 4–5 × 1–1.5 microns.

Edibility   Inedible because too tough.

Habitat   Solitary or imbricate on dead wood of both conifers and broadleaf trees, widely distributed.

Season   Spring to late autumn.

Note   *T. chioneus* (*T. albellus*) is similar but whitish and very common on dead broadleaf wood.

309   POLYPORUS SQUAMOSUS

Etymology   From Latin, "scaled."
Description   Cap 10–60 cm, ochreous, variegated with adpressed brown scales, fan-shaped or hemispherical, convex,
flat then concave, often umbilicate when young, margin thin,
cuticle dry. Tubes 0.2–1 cm long, decurrent, not detachable,
cream-colored; pores quite large, whitish then pale yellow, angular. Stipe 3–8 × 1–5 cm, central or more usually lateral,
solid, sturdy, squat, ochreous, reticulate uppermost, blackish
at base. Flesh white, soft, thick, then leathery-suberose.
Strong odor, cheeselike then mealy, with flavor of watermelon
rind. Spores white, elliptical, smooth, 10–12 × 4–5 microns.
Edibility   While edible, this is not to everyone's liking because of its flavor. Some people deep-fry the tender edges,
some pickle them, and others boil the mushroom to add flavor
to a soup stock and then discard the boiled mushroom.
Habitat   Parasitic on various broadleaf species.
Season   Spring through autumn.
Note   This fungus is fairly abundant in eastern spring woods,
especially on elms.

---

310   ISCHNODERMA BENZOINUM

Synonym   *Polyporus resinosus.*
Etymology   After its odor which is like benzoin.
Description   Carpophore 8–20 cm wide, 1–2.5 cm thick,
bracket- or oyster-shaped, sessile, but attached with a tuberculate base, surface blackish brown with bright blackish blue
highlights, rugose, downy, often with droplets of resinous exudate; margin thin in mature specimens, often lobate, brown.
Tubes thin, whitish then brown; pores roundish, whitish tinged
with rust-brown, yellowish, finally tobacco-brown. Flesh fibrous
then suberose-ligneous, first whitish then rust-brown. Changeable, sometimes aniseedlike odor. Spores white, cylindrical-arcuate, smooth, 4–7 × 1.5–2.5 microns.
Edibility   Inedible because too fibrous.
Habitat   Fairly rare but found in all cold-temperate and cold
regions of the Northern Hemisphere, on dead coniferous trunks
and stumps; sometimes a not very active parasite.
Season   Spring to late autumn, with carpophores persisting
on wood.
Note   Some mycologists recognize a distinct species, *I. res-
inosum,* which lacks bluish tones and grows on broadleaf
trees; others consider it only a variant.

311   OSMOPORUS ODORATUS

Synonym   *Trametes odorata.*
Etymology   From Latin, "scented."
Description   Carpophore 6–5 × 2–8 cm, bracket-shaped, sessile; or turbinate-circular when growing on a horizontal surface, yellowish-brown or reddish, or older parts at back blackish; quite irregular with conspicuous nodular excrescences, divided into broad braceletlike forms by deep, close grooves, very downy, rugose, margin thick, obtuse, lobate, whitish or more often bright yellow or also orange, finally rust-brown. Tubes yellowish, not perfectly stratified; pores roundish, distant, feltlike, yellowish then ochreous rust colored. Flesh orange-brown, suberose, first soft then hardening. Strong smell of aniseed or vanilla tending to disappear with age. Spores white then darkening, elliptical, smooth, 6–9 × 3–5 microns.
Edibility   Inedible.
Habitat   On coniferous wood, especially Norway spruce stumps, also on beams, where they form aberrant carpophores.
Season   All year.
Note   The strong odor of the carpophores is also present in wood invaded by the hyphae, causing a breakdown of the tissue ("rot"), which turns red.

312   ECHINODONTIUM TINCTORIUM

Common name   Indian paint fungus.
Etymology   From Latin, "used for dyeing."
Description   Carpophores 5–30 cm, bracket-shaped, hoof-shaped, sessile, upper surface black and cracked. Hymenium in early stages and toward edge with pores or mazelike tubes, gradually changing into teeth 1–2 cm long, varying in color, eventually grayish then black. Flesh ligneous, leathery, brick-red. Spores whitish, oval, minutely spiny, 6–8 × 3–5 microns.
Edibility   Inedible because too woody.
Habitat   Parasitic on mature conifers in western North America.
Season   All year, growing from spring to autumn.
Note   This is a common and widespread parasite on conifers, causing tremendous damage to the forests of North America. It is in fact a living fossil of the fungus kingdom, the only species which forms ligneous and perennial carpophores with hymenium of teeth. When ground up and reduced to fine powder the flesh of this fungus was used by the indigenous peoples to paint war signs on their bodies, and for other purposes. The same powder was also used by shamans as medicine; in fact it contains various alkaloids and tannin.

313 SPONGIPELLIS PACHYODON

Synonyms *Irpex pachyodon; Irpex mollis.*
Etymology From Greek, "large-toothed," because of the shape of the hymenium.
Description Carpophore 3–8 cm, bracket-shaped or fairly crustaceous, sessile, glabrous uppermost, white then cream-colored with darker striations. Hymenium cap-colored tinged with pink, variable, formed by teeth 1–1.5 cm long, fairly joined together forming sinuate canaliculi or false gills, especially at the front. Flesh rather leathery, white, unchanging. Spores white, subglobose or broadly elliptical, smooth, 5–8 × 4.5–6.5 microns.
Edibility Inedible because of the leathery texture.
Habitat Parasitic on various broadleaf species.
Season All year.
Note Many species of *Trametes*, a thin bracket-shaped fungus with a hymenium with pores, may form carpophores with separating tubes that resemble the toothlike projections of *S. pachyodon.* A similar but more crustlike white polypore is *Irpex lacteus.*

314 HERICIUM ERINACEUS

Synonyms *Dryodon erinaceus; Hydnum erinaceum.*
Etymology From Latin, "porcupine," from its appearance.
Description Carpophore 5–30 cm, spatula-shaped, formed by thick, interwoven branches, often partly joined together, with a rudimentary stipe, white tending to yellow, covered by teeth 3–6 cm long, slender, pendant, supple, pruinous. Flesh white, unchanging, thick and cavernous, soft but slightly elastic. Odor slightly acid, flavor sweetish. Spores white, subglobose, smooth, 6–7 microns.
Edibility Good.
Habitat Parasitic on various broadleaf trees.
Season Late summer to early spring.
Note The dried and ground flesh is used as a styptic agent by some people. *H. coralloides,* which may also grow on conifers, has distinct clusters of shorter teeth (2 cm at most), and its flesh turns blue with iodized solutions.

315 STECCHERINUM OCHRACEUM

Etymology From Latin, "ochre-colored."

Description Carpophore 2–7.5 cm, bracket-shaped, mainly adherent to the substratum, often crustaceous, sessile, semi-circular, then merging with other fungi, margin undulate-sin-uate, white, membranous and pubescent; sterile surface zonate, downy, whitish or pale yellow-ochre. Hymenium with ochreous teeth, tinged with bright yellow-pinkish, small, shorter at the margin. Flesh thin, membranous, leathery, white. Odor and flavor negligible. Spores white, ovoid, smooth, 3–4 × 2–2.5 microns.

Edibility Of no interest because of texture.

Habitat On dead branches or on the bark of living trees.

Season All year.

Note The carpophore of many lignicolous species, depending on the substratum and its structure, may grow either as small brackets (half-carpophores) or as crusts (resupinate carpophores).

316 STEREUM HIRSUTUM

Etymology From Latin, "covered with hair."

Description Carpophore 2–10 cm, yellowish or gray with yellow margin, broadly adherent to the substratum, sometimes crustaceous, sessile, upper part bracket-shaped, covered with evident hairs, slightly zonate. Hymenium smooth, bright ochreous or rarely pinkish, leather-colored, grayish. Flesh thin, membranous, leathery. No odor or flavor. Spores white, elliptical, curved, smooth, 5–8 × 2–4 microns.

Edibility Inedible because too tough.

Habitat On the wood of dead broadleaf species.

Season All year.

Note This is a fairly common lignicolous species, easy to spot because of the color of the hymenial surface. *Stereum* is a genus of crustlike or bracket-shaped fungi that grow on wood and resemble several thin-fleshed polypores except that their hymenium is smooth instead of porelike.

317 STEREUM INSIGNITUM

Etymology From Latin, "friezed" or "ornate."
Description Carpophores 2–10 cm, bracket-shaped and semicircular or concrescent-lobate, sessile, normally imbricate, with rust-brown velvety zonations uppermost, then glabrous, fibrous, wine-brown with age, alternating with lighter, concentric, yellowish brown, downy zones, sometimes greenish because of algae, margin undulate and lobate, ochreous yellow. Hymenium smooth, pale ochre, sometimes tinged with pink, finally pale gray. Flesh thin, leathery with a lower whitish stratum and a shiny reddish brown upper stratum. No particular odor or flavor. Spores whitish, elliptical, smooth, 5.6–6.5 × 2.5–3 microns.
Edibility Inedible because too tough.
Habitat On oak and broadleaf trunks, in warm temperate regions.
Season All year, growing from spring to autumn.
Note *S. subtomentosum*, though quite similar, is easily identifiable by the fact that the hymenium turns chrome-yellow when touched. It grows on various broadleaf species, mainly poplar and willow. Most of the other species which may cause confusion have the sterile surface with markedly villose zonations, particularly towards the edge. Neither has been found in North America.

---

318 CHONDROSTEREUM PURPUREUM

Synonym *Stereum purpureum.*
Etymology From Latin, "purple."
Description Carpophore 2–6 cm, crustaceous or adpressed to substratum, sessile, margin free and bracket-shaped, connate and imbricate, sterile surface feltlike and downy, grayish-white to brownish, with slight zonations, margin acute, undulate, sinuate. Hymenium smooth, waxy in damp weather, pruinous when dry, initially whitish then purple-violet or pale lilac-brown, finally dark-brown or violet, tending to fade. Flesh thin, leathery-soft, horny in dry state, with various distinct strata, whitish. Spores white, elliptical and ovoid, smooth, 6–8 × 3–4 microns.
Edibility Of no interest because thin and leathery.
Habitat Worldwide on broadleaf trees, rarely on conifers.
Season All year.
Note The parasitism of this fungus causes a withering of the upper part of the leaf lamina.

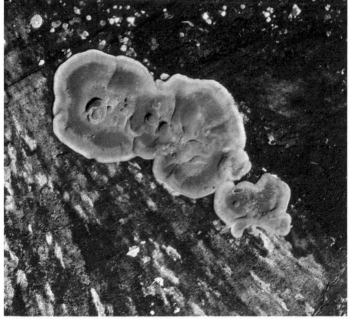

319 PULCHERRICIUM CAERULEUM

Synonym *Corticium caeruleum.*
Etymology From Latin "(sky-) blue."
Description Carpophore 2–5 cm, prone, sessile, at first rounded, dark blue, downy with edges normally lighter, then merging with others in sheets. Hymenium faintly rugose or tuberculate. Flesh thin, membranous, soft, fairly detachable when fresh, more adherent to substratum when dry, bluish. No odor or flavor. Spores white, oval, smooth, 7–11 × 5–7 microns.
Edibility Of no value because of texture.
Habitat On dead wood in warm temperate zones.
Season All year.
Note This is a species of no gastronomic interest but conspicuous because of its beautiful color. It is most common in the Southeast, where it covers fallen oak branches with a deep blue felt.

---

320 TREMELLA MESENTERICA

Etymology From Greek, "like the middle intestine."
Description Carpophore 1–8 cm, orange-yellow, variously contorted, plicate, undulate, circumvolute, cerebriform, pruinous because of the spores. Flesh gelatinous, yellow, becoming elastic when drying, then hard. Insignificant odor and flavor. Spores white, elliptical, smooth, 13–14 × 7–8 microns.
Edibility Of no interest because of texture.
Habitat On dead wood, especially small branches.
Season All year.
Note This fungus can be very common in the spring and autumn. A pale yellow form that is more lobed than brainlike is *T. lutescens; T. foliacea* is leaflike and reddish brown, can be quite large; and *T. reticulata* is creamy and coral-like. The closest look-alike is *Dacrymyces palmatus,* which is brainlike and orange-yellow, but has a white point of attachment to the wood and is very different microscopically.

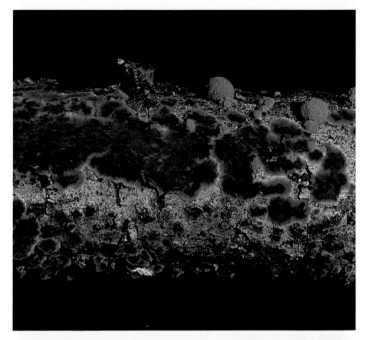

321 AURICULARIA AURICULA

Common names Tree-ear; wood-ear.

Etymology From Latin, "ear."

Description Carpophore 6–10 cm, ear-shaped, sessile or with a short attaching peduncle, outer surface sterile, pubescent, with slight venations; inner surface fertile, reddish brown, at first almost smooth then venose, pruinous because of the spores. When drying out it tends to turn violet and become increasingly circumvolute. Flesh soft, gelatinous, slightly elastic, translucent, fragile when dry, reviviscent. No particular odor or flavor. Spores white, cylindrical, smooth, 12–17 × 4–7 microns.

Edibility Good, although its value is more in its texture than its flavor.

Habitat On broadleaf wood.

Season All year, especially in rainy periods, in winter and spring.

Note A very similar species, *A. polytricha,* is grown commercially in the Orient, and sold here in Oriental groceries; it is called *Mu ehr,* "small ear." Besides being eaten in soups and vegetable dishes, it is used in folk medicine to soothe coughs and generally improve the physical condition. *A. auricula,* however, is not known to confer any medical benefits.

322 AURICULARIA MESENTERICA

Etymology From Greek, "intestinelike," from its shape.

Description Carpophore 5–12 cm, bracket-shaped, sessile, edge lobate, gray-brown, outer surface sterile with whitish zonations, quite villose, pale blue-gray, slightly velvety; lower surface fertile, very pleated, purple-brown, pruinous because of the spores. Flesh gelatinous, quite thick and elastic, leathery when dry. No particular odor or flavor. Spores white, slightly incurved, smooth, 15–18 × 7–8 microns.

Edibility Inedible because of elastic texture.

Habitat In groups on broadleaf wood.

Season All year, especially in winter and spring.

Note In the photograph *A. mesenterica* is on the right; a polypore, probably *Trametes versicolor* (**300**), is on the left.

323 DACRYMYCES DELIQUESCENS

Etymology From Latin, "dissolving."
Description Carpophore 0.2–1 cm, lentiform, sometimes roundish, yellow or orange-yellow, with no margin, finally contorted, usually sessile, sometimes with a very small stipe. Flesh pale yellow, transparent, gelatinous. No odor or flavor. Spores white, cylindrical, incurved, septate, smooth, 8–22 × 4–7 microns.
Edibility Of no interest because of its texture.
Habitat Gregarious on dead wood and fallen branches, usually conifers.
Season All year, more frequent in cool weather.
Note This belongs to a small group of gelatinous fungi which grow parasitically and saprophytically on wood. *D. palmatus* is brain-shaped and reddish orange. Other cuplike to cushion-shaped jelly fungi, also some shade of yellow, include species of *Femsjonia* and *Guepiniopsis*.

324 EXIDIA GLANDULOSA

Etymology From Latin, "full of glands," because of the shape of the hymenium.
Description Carpophore 5–10 cm wide, blackish, globose or lenticular, attached by a tiny peduncle, sometimes flat, undulate, tumorlike, becoming ash-colored or faintly downy; when drying out it forms a black crust on the substratum. The fertile upper part is covered with numerous small conical papillae. Flesh blackish, gelatinous, transparent, soft. No odor or flavor. Spores white, ovoid or cylindrical, sometimes incurved, smooth, 12–15 × 4–5 microns.
Edibility Not known.
Habitat On dead wood, especially broadleaf branches.
Season Autumn to spring.
Note Other members of this genus also grow on dead wood: *E. truncata*, brownish, hemispherical with a short peduncle; *E. nucleata*, glassy or tinged with lilac, finally brick-colored, at most 1 cm wide; *E. thuretiana*, opal or pinkish, drying out into a yellow crust; and *E. viscosa*, white, grayish or violet, completely smooth and quite viscid.

325 GRIFOLA FRONDOSA

Common name Hen-of-the-woods.
Synonym *Polyporus frondosus.*
Etymology From Latin, "covered with leaves."
Description Carpophore 20–40 cm in diameter, bushlike, hemispherical, made up of many imbricate caps, very crowded in young specimens, spatula-shaped, up to 10 cm wide, undulate, brown, darker with faint zonation toward margin, velvety, covered with small clusters of radial fibrils; edge covered with pores; stipes connate and merging into a sturdy stipe. Tubes decurrent to branch attachments, difficult to detach, up to 3 mm long, whitish; pores round or mazelike, white. Flesh whitish, quite fibrous, slightly watery. Odor mushroomy with a hint of meal, flavor sweet. Spores white, subelliptical, smooth, 5–7 × 3.5–4.5 microns.
Edibility Choice, very sought after in some areas.
Habitat Parasitic on the roots of broadleaf trees, particularly oak.
Season Throughout the autumn.
Note This is one of the largest fungi (it can weigh more than 25 pounds) and one of the most popular autumn edibles. It somewhat resembles a small hen, is often found covered with leaves, and grows at the base of living oaks.

326 MERIPILUS GIGANTEUS

Synonym *Polyporus giganteus.*
Etymology From Latin, "gigantic."
Description Carpophore with bracket-, spatula-, or fan-shaped caps, imbricate or superposed, 10–30 cm wide, brown, pale then yellowish brown at margin, blackish brown at rear and cream or chamois-colored at margin, velvety or granulose, with faint zonation, radially sulcate, depressed at rear toward base. Hymenium with whitish, decurrent tubes, soot-brown then blackish when touched, 1–2 mm long; pores small, round or polygonal, whitish, turning blackish when touched. Flesh white, darkening, tough, quite leathery, fibrous. Odor and flavor acidic. Spores white, globose, smooth, 4–5 microns.
Edibility Good when young, but slightly leathery.
Habitat In crowded groups on broadleaf stumps, usually beech or oak, often at the foot of living trees which are in poor health.
Season Late spring to early winter.
Note Has a marked degrading effect on wood; sometimes produces an extensive collection of carpophores. Up to 180 pounds of fungi have been gathered from old chestnut trunks in Europe.

327 POLYPORUS UMBELLATUS

Etymology From Latin, "with umbrellas."

Description Carpophore bushlike (up to 50 cm in diameter), consisting of caps of 1–4 cm, convex then umbilicate with undulate and sinuate margin, fairly light brown, radially fibrillose, pruinous or slightly squamose, each on a branch. Stipe white or reddish brown, often originating from a large hypogeal sclerotium, divided into numerous pruinous, cylindrical ramifications. Tubes very short, decurrent on branches, reduced sometimes to mere alveoli; pores wide, irregular, white. Flesh white, soft and juicy, slightly fibrous. Odor slightly mealy, flavor sweet. Spores white, elliptical, smooth, 7–10 × 3–4 microns.

Edibility Choice.

Habitat At the foot of old broadleaf species, probably also parasitic on them.

Season Spring to autumn.

Note Although uncommon, where this fungus does occur it will often appear year after year, and its clusters are usually large and provide many meals. This is one of very few polypores that are both tender enough to eat and that taste good.

---

328 THELEPHORA TERRESTRIS

Etymology From Latin, "growing on the ground."

Description Carpophore consists of several laminae, bent or conchoidal, 3–5 cm high, becoming crustaceous on the substratum at base and merging laterally, imbricate uppermost with rust-brown fibrillose scales, blackish brown with age, frayed with white at margin then same color as the rest. Hymenium rugose or papillate, chocolate-brown and pruinous. Flesh dark brown, leathery but soft, fibrillose. Odor of damp loam, flavor quite astringent. Spores dark brown, globose-angular, warty, 8–9 × 6–8 microns.

Edibility Inedible because acrid and too leathery.

Habitat On the ground in sandy pine forests, on roots and twigs.

Season All year, usually in late autumn.

Note The species belonging to this genus usually grow in bushlike formations of varying shapes: e.g., *T. palmata,* with laminae divided at the margin into compressed thimble-shaped sectors; *T. anthocephala,* with pedunculate laminae; *T. clavularis,* with pointed, subcylindrical branches; and *T. caryophyllea,* with often finely incised lobate laminae.

329 SPARASSIS CRISPA

Etymology From Latin, "curled."
Description Carpophore 10–60 cm, like a cauliflower, whitish or slightly ochreous, evidently ramified, with flattened, broad, ribbonlike and intricate branches, apex yellowish and curled. Stipe white, darkening with age, squat, rooting. Flesh whitish or faintly pale yellow, fragile. Odor quite pleasant, faintly of aniseed, flavor pleasant, of walnut. Spores whitish to light ochre, ovoid, smooth, 6–7 × 4–5 microns.
Edibility Choice.
Habitat On coniferous wood, at the foot of trees or on roots.
Season Autumn.
Note This may be the same as the western *S. radicata*, which has a large rooting stipe. The common eastern species, commonly called *S. crispa*, should probably be called *S. herbstii*; it grows on the ground near trees or on very rotten wood in broadleaf woodlands.

330 RAMARIA FORMOSA

Synonym *Clavaria formosa.*
Etymology From Latin, "handsome."
Description Carpophore up to 30 cm high and 15 cm wide, bush-shaped, with a thick, solid, rooting stipe, sometimes already divided into several branches from the base, white, faintly pinkish; ramifications crowded, cylindrical, rugose lengthwise, pinkish buff to salmon-colored, with yellowish tips, ochreous and powdery because of the spores when mature. Flesh white, darkening slowly when exposed to air, thin, pinkish-gray with aniline water. Odor insignificant, flavor slightly bitter. Spores ochreous, elliptical, rugose, 12–15 × 4–5 microns.
Edibility Inedible, a powerful purgative.
Habitat On humus-rich ground beneath broadleaf trees.
Season Summer to autumn.
Note There are a great many species of branched coral fungi, and they are very difficult to tell apart. Because *R. formosa* is known to cause gastrointestinal distress, and because some others are somewhat indigestible, it is best to proceed with great caution when experimenting with any species of *Ramaria*.

331 RAMARIA FLAVA

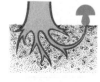

Synonym *Clavaria flava.*

Etymology From Latin, "yellow."

Description Carpophore at most 20 cm high and 15 cm wide, bushlike. Stipe thick, short, narrowing at the base, solid and white; branches straight, quite distant, cylindrical or often compressed, divided several times at the bottom, dichotomous at the top, sulfur-yellow or lemon-yellow, ochreous when mature, tips branch-colored, denticulate. Flesh white, marbled, tender. Light odor, flavor sweet and pleasant. Spores ochre, elliptical, rugose, 9.5–15.5 × 3.2–6.5 microns.

Edibility Good, but avoid older specimens.

Habitat In broadleaf and coniferous woodland.

Season Summer to autumn.

Note *R. flava* actually is a complex of yellowish coral fungi. The species differ somewhat in color, branching, and microscopic characters. None is known to be poisonous but caution is recommended because of the somewhat similar *R. formosa* (330).

---

332 CLAVICORONA PYXIDATA

Synonym *Clavaria pyxidata.*

Etymology From Latin, "with small vases," because of the form of the apical branchlets.

Description Carpophore 5–12 cm, whitish or faintly hazel, sometimes darkening when mature. Stipe thin, smooth, quickly ramifying; branchlets like upturned cones, concave, almost hollow in the apical cup, surmounted by teeth. Flesh white, solid. Faint odor, flavor slightly peppery. Spores whitish, elliptical, smooth, 4 × 3 microns.

Edibility Can be eaten, but is a slight laxative.

Habitat On rotting wood.

Season Late summer and autumn.

Note This is one of the few coral fungi that grow on wood. It is further recognized by the candelabralike clusters of branching tips.

333 RAMARIA BOTRYTIS

Synonym *Clavaria botrytis.*
Etymology From Greek, "like a bunch of grapes."
Description Carpophore up to 15 cm high and 20 cm wide, bush-shaped. Stipe stout, narrowing at the base, white, pale yellow when mature, divided into short cylindrical branches, white then ochreous, surmounted by short, denticulate tips, coral-red or brick-red at the apex, ochreous when mature. Flesh firm, brittle, white, unchanging. Odor slightly fruity, flavor sweetish. Spores ochre, fusiform, rugose or subreticulate, 12–17 × 4–6 microns.
Edibility Good.
Habitat On the ground under conifers.
Season Summer and late autumn.
Note Species of *Ramaria* are generally recognized by spore size and shape or other microscopic or chemical characteristics. Field identification is rarely reliable for species of this genus.

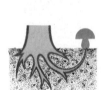

334 RAMARIA INVALII

Synonym *Clavaria invalii.*
Etymology After the town of Inval, Surrey, England.
Description Carpophore 4–5 cm high, golden yellow, bush-like, compact, dense, almost spherical. Stipe fairly evident, short, often hairy because of white or yellowish rootlike filaments, pointed; numerous branchlets, thin, short, unequal, straight, cylindrical, smooth, with tipped apex. Flesh white. Odor slightly sharp, flavor slightly bitter. Spores ochreous yellow, spinelike, 7–9 × 4 microns.
Edibility Unknown.
Habitat In coniferous litter under western hemlocks.
Season Summer and autumn.
Note Other small species of *Ramaria* grow in the humus in coniferous woodland, such as *R. ochraceovirens*, olive-yellow turning green when rubbed; *R. palmata,* very pale ochre with compressed branchlets; and *R. gracilis,* pale orange with whitish upper branches, and sometimes with a strong odor of aniseed.

335 RAMARIA MAIREI

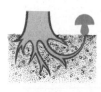

Synonym *Clavaria pallida.*
Etymology From Latin, "discolored."
Description Carpophore 6–18 cm high, bushlike. Stipe thick, short, café-au-lait in color, whitish at base, with crowded, dichotomous branchlets, rugose lengthwise, first whitish then tinged with pink, then ochreous as well; tips obtuse with short teeth, pale purple when young. Flesh white, not hygrophanous. Slight soaplike odor, slightly bitter aftertaste. Spores ochreous, elliptical, spotted, 9–12 × 5–6 microns.
Edibility Normally causes violent and painful intestinal disorders; becomes bitter when cooked.
Habitat In woodland, preferably beneath fir.
Season Summer and autumn.
Note Although there are no typical specimens (and it has not been reported in North America), this fungus is identifiable by the bifurcations of the evidently rounded branchlets. *R. testaceoflava* is at most 5 cm high, stipe reddish, branchlets cinnamon-red, tips yellow; the flesh is wine-red when exposed to the air; grows under western hemlocks. *R. fennica* is also purple or violet when young. *R. fennica violaceibrunnea*, which grows under western conifers, turns brownish when mature.

336 CLAVARIA INAEQUALIS

Etymology From Latin, "unequal."
Description Carpophore 2.5–6 cm, golden yellow, pale sulfur-yellow at the base, usually single, cylindrical, tip pointed, sometimes forked, solid. Flesh pale yellow. No particular odor or flavor. Spores white, globose or ovoid, smooth, 7–9 × 6–8 microns.
Edibility Of no interest.
Habitat In woodland, open places, and heathland.
Season Summer and autumn.
Note *C. inequalis* is an ambiguous name. Different mycologists have described different entities as *C. inaequalis,* and the diagnostic differences among these otherwise identical species are microscopic. The very common *C. fusiformis* is similar, with typically unbranched, yellow, tubular to compressed clubs that are more or less joined at the base. *C. vermicularis* is similar but white, and *C. purpurea* is purplish brown. None is known to be poisonous and *C. fusiformis,* when found in quantity, is worth gathering to add to a cooked vegetable dish.

337 CLAVULINA RUGOSA

Synonym   *Clavaria rugosa.*
Etymology   From Latin, "wrinkled."
Description   Carpophores 5–12 cm high, white, single or slightly ramified, enlarged at the top, up to 1 cm wide, irregular and rugose, tuberculate, sulcate. Ramifications few in number, irregular, obtuse but rarely cristate. Flesh white, abundant, firm. No particular odor or flavor. Spores white, subglobose, smooth, 8–9 × 6–8 microns.
Edibility   Can be eaten.
Habitat   Gregarious in woodland and adjacent grassy areas.
Season   Early summer to late autumn.
Note   Although quite variable, it is typically only slightly branched, whitish, and wrinkled. In addition, it has only two-spored basidia with strongly incurved sterigmata; these characteristics, while microscopic, are useful for precise identification. Other species with similar microscopic characteristics include the more densely branched, whitish *C. cristata* and ash-gray *C. cinerea.*

338 CLAVARIADELPHUS PISTILLARIS

Synonym   *Clavaria pistillaris.*
Etymology   From Latin, "pestle-shaped."
Description   Carpophore 8–25 cm high with a maximum diameter of 2.5 cm, club-shaped, slender or stout, surface at first smooth then sulcate lengthwise, light yellow tending to become grayish brown. Flesh white, bruising brownish, solid, soft, then spongy-fibrous. Odor pleasant, flavor somewhat bitter. Spores cream-colored, elliptical, smooth, 9–16 × 5–7 microns.
Edibility   Edible but quite poor in flavor.
Habitat   Beneath broadleaf species, particularly beech, in calcareous ground.
Season   Summer to autumn.
Note   This is a gregarious and quite common species. This genus includes other club-shaped fungi such as *C. truncatus* (**339**); *C. ligula,* up to 10 cm high, with a hairy base, light ochre then reddish, grows beneath conifers; *C. fistulosus,* up to 20 cm high, very thin, yellowish, hollow, with a white and downy base which grows on fallen branches, including the stems of leguminous plants; and *C. junceus,* up to 15 cm high, very thin, at most 2 mm in diameter, ochreous and growing on the litter in broadleaf woods.

339 CLAVARIADELPHUS TRUNCATUS

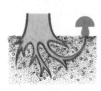

Synonym *Clavaria truncata.*
Etymology From Latin, "cut or broken off."
Description Carpophore 5–15 cm high, conical, turbinate, with the upper part convex then flattened, slightly depressed, irregular, rugose, light yellow then bright yellow or orange, margin sulcate and lobate, very fragile, sterile like the whole upper surface, 3–8 cm wide. Lower part is like a stipe which tapers conspicuously toward base, mainly fertile, at first smooth then sulcate lengthwise or with small ramified veins at top, dirty pale lilac, tending to turn yellow with the spores, yellowish at edge. Stipe feltlike and white. Flesh whitish, reddish when rubbed, thick and cottony. Faint odor and a sweet, sharply sugary flavor. Spores pale yellow, ellipsoid, smooth, 10–11 × 6–7 microns.
Edibility Edible, but slightly laxative in quantity.
Habitat Gregarious in mountains in coniferous woods.
Season Summer and autumn.
Note The sugary flavor is due to the large amount of mannitol, a sweet-tasting higher alcohol found in the exudate of certain plants. It is produced commercially from certain species of maple and ash, and is used as a mild purgative for infants, in the production of alcoholic beverages, and in the pharmaceutical industry.

340 CALOCERA CORNEA

Etymology From Latin, "hornlike," because of its appearance.
Description Carpophore up to 1 cm high, cylindrical, pale yellow, white and hairy at the base, tapering at the top, pointed, isolated or in small clusters. Flesh virtually nonexistent, yellow, and elastic. Spores ochreous, elliptical, depressed laterally, smooth, 7–9 × 4–4.5 microns.
Edibility Of no value because of size.
Habitat On fallen broadleaf and coniferous trees.
Season All year.
Note Although this may resemble some kind of small unbranched coral or tooth fungus, its texture is gelatinous-horny and it is related to those jelly fungi that possess tuning-fork-shaped basidia.

341 CALOCERA VISCOSA

Etymology From Latin, "sticky."
Description Carpophore 3–6 cm high, ramified, bright yellow, orange-yellow, each branchlet ending with two or three obtuse tips, very short; surface barely viscid. Stipe dry, hairy, extending into a white mycelial thread. Flesh yellow, tough, elastic. Spores bright yellow-ochre, elliptical, smooth, 7–12 × 4–4.5 microns.
Edibility Of no interest because of texture.
Habitat On rotting coniferous wood.
Season All year, more frequent in summer and autumn.
Note This resembles a slightly branched coral mushroom but, where corals are brittle, this is somewhat rubbery and horny-hard. Jelly fungi are usually found on wood and vary in texture from soft and jellylike to rubbery-tough. In appearance they resemble fungi in several distinct orders including cup fungi, corals, and tooth fungi.

342 GYMNOSPORANGIUM CLAVARIAE-FORME

Etymology From Latin, "clavarialike"
Description Carpophore 1–2 × 0.2–0.3 cm, shaped like small clubs, cylindrical, compressed, sometimes divided in the top part, orange-yellow, gelatinous. Spores of two types, thick- or thin-walled, ochreous, elliptical, smooth, with one septum and a long penduncle; 76–86 × 13–17 microns in the first instance, 80–105 × 13–15 microns in the second.
Edibility Undetermined.
Habitat On juniper branches, mainly *Juniperus communis* (common juniper).
Season Spring.
Note This is a "rust" which lives as a parasite on two different host plants. In the spring it lives on juniper, where it forms swellings like tumors; in summer, on the lower pagina of various Rosaceae, particularly the service tree and hawthorn, where it forms yellow reproductive structures which are enlarged with thimble-shaped extensions, known as *Roesetelia lacerata*. Later, the infection is transmitted alternately from juniper to service tree, for example, and vice versa. More common is the cedar apple rust, *Gymnosporangium juniperi-virginianae*, which is very similar in development and appearance; it occurs alternately on *Juniperus virginiana* (eastern red cedar) and on apple and crabapple.

343 MICROGLOSSUM VIRIDE

Common name Green earth tongue.

Etymology From Latin, "green," from its color.

Description Carpophore 3-9 cm high, club-shaped, dark green, olive, compressed or sulcate lengthwise, glabrous, but slightly viscous in damp weather. Stipe shorter than the upper fertile part, cylindrical, finely squamulose. Odor and flavor negligible. Spores whitish, fusiform, smooth, 12-17 × 5 microns.

Edibility Of no interest because of size and rarity.

Season Late summer and fall.

Habitat On the ground in grass or woods in eastern North America.

Note *M. olivaceum* is olive to olive-brown; the more common *M. rufum* is similar in shape but yellow to orange in color. The genus *Microglossum* resembles an unbranched coral fungus but it is an Ascomycete, producing its spores in spore sacs (asci) rather than on the end of club-shaped structures (basidia) as do Basidiomycetes. Other earth tongues include the blackish genera *Trichoglossum* and *Geoglossum*, and the irregularly shaped, yellow *Neolecta irregularis*.

344 XYLARIA HYPOXYLON

Etymology From Greek, "almost ligneous or wooden," because of the texture.

Description Stromata (the cushionlike masses in which spores are produced) up to 8 cm high, often branched. Stipe cylindrical, black, the upper part white, powdered because of the asexual conidiospores; or enlarged and unbranched as a tapering, pointed cylinder, without conidiospores, but papillate because of the immersed perithecia. Spores black, bean-shaped, smooth, 11-14 × 5-6 microns.

Edibility Of no interest because of texture.

Habitat On dead wood.

Season Spring and autumn, but can appear at any time where conditions are favorable.

Note While this species typically occurs on rotten wood, variants or similar species can be found on cones, magnolia fruits, hawthorn berries, and leaves.

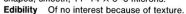

345 XYLARIA POLYMORPHA

Common name   Dead man's fingers.
Etymology   From Greek, "of many shapes," because of its variability.
Description   Stromata up to 9 × 2.5 cm, quite clublike, with a short cylindrical stipe, often slightly lobate in the upper fertile part, pointed at the tip, solitary or in small groups, blackish, opaque, densely curled or minutely cracked around the papillae corresponding to the perithecia. Flesh hard, tough, white with a blackish layer of ovoid perithecia beneath the black superficial crust. Odor and flavor negligible. Spores dark brown or blackish, fusiform, smooth, 20–34 × 5–9 microns.
Edibility   Of no interest because of the consistency.
Habitat   On rotting wood, especially beech, with stromata emerging at ground level.
Season   Present all year, but formed in summer and autumn.
Note   The young stromata, as they form, are covered with ochre-brown asexual spores (conidia). It gets its common name from the appearance of the generally clustered clublike structures that rise out of the ground or from rotting wood about ground level.

---

346 CLAVICEPS PURPUREA

Common names   Ergot; spurred rye.
Etymology   From Latin, "purple."
Description   Sclerotium elongated, fairly cylindrical, with rounded ends, slightly grooved, growing on the ovary of graminaceous flowers and eventually replacing it. It varies quite a lot in shape and size in relation to the plant infected. Outwardly blackish, internally white and hard. The stromata emerge, either isolated or in small groups, from the sclerotium when it drops to the ground; they consist of a globose head, about 2 mm in diameter, ochreous brown or pale purple, spotted with immersed perithecia and a long, thin, smooth head-colored stipe. Spores colorless, filiform, septate, smooth, 100 × 1 micron.
Edibility   Extremely poisonous.
Habitat   Sclerotia parasitic on the inflorescences of Graminaceae, especially rye.
Season   The sclerotium grows over summer in the grass flower, falls, overwinters, and produces small mushroomlike fruit-bodies the following spring; the spores from these fruit-bodies infect the grass flowers and become that summer's sclerotia.
Note   If the sclerotium is ground up with rye, barley, or wheat, it causes violent poisoning known as ergotism or St. Anthony's Fire.

347 CORDYCEPS MILITARIS

Etymology From Latin, "military," because its shape resembles a plume.

Description Stromata up to 5 cm high, club-shaped, normally solitary, orange-red. Stipe supple, imperceptibly extending into the fertile cylindrical head, which is up to 5 mm wide and finely spotted by immersed perithecia. Spores colorless, filiform, smooth, breaking up into secondary, rodlike spores, 3.5–6 × 1.5 microns.

Edibility Of no interest.

Habitat On lightly buried butterfly larvae and pupae, especially common in the Southeast.

Season Late summer and autumn.

Note The stromata of this species contain a powerful antibiotic which stops their alteration by bacterial degradation. Another species which is parasitic on insects, *C. sinensis,* used as an antityphoid agent in Chinese folk medicine, has also shown itself to be rich in antibiotic compounds.

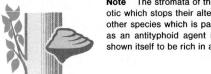

348 CORDYCEPS MEMORABILIS

Etymology From Latin, "memorable."

Description Stromata up to 1.5 cm long and 0.1–0.3 mm wide, supple, depressed, sulcate lengthwise, orange, rust-colored, minutely furfuraceous, single, some concrescent at the base and sometimes joined lengthwise with ovoid, bare, completely free perithecia, at most 0.5 mm high, variously positioned, grouped in rows or scattered, smooth, rust-colored, apex sepia-brown. Spores colorless, filiform, smooth, breaking up into secondary elliptical spores, 1.7–2 × 1.3 microns.

Edibility Of no interest.

Habitat On larvae and adult members of Coleoptera (beetles and weevils).

Season Midsummer.

Note It represents the sexual and fairly rare form of a common mold that parasitizes insects, *Paecilomyces farinosus,* which may also form whitish, elongated, sporigenic structures called coremia. The formation of the coremia is stimulated by sunlight.

349 CORDYCEPS CAPITATA

Etymology From Latin, ''with a head.''
Description Stromata up to 10 cm high, solitary or in small groups. Stipe quite stout, about 1 cm wide at the base, yellow, smooth or sulcate, suddenly enlarging into an ovoid or subglobose head, dark brown or blackish, densely roughened by the perithecial ostioles, which are immersed, particularly when the stroma dries out. Spores colorless, filiform, smooth, breaking up into secondary cylindrical spores, 7–20 × 2–3 microns.
Edibility Of no interest.
Habitat Parasitic on the hypogeal fruit-bodies of *Elaphomyces*.
Season Summer and autumn, like their host carpophores.
Note This can be rather common in eastern fall woods. The false truffle that it grows out of once was thought to be an aphrodisiac, and was sold in English apothecaries. Because of the shape of the *Cordyceps* attached to the false truffle this has been used as a fertility symbol in various rites in Mexico.

350 CORDYCEPS OPHIOGLOSSOIDES

Etymology From Greek, ''like a snake's tongue.''
Description Stromata up to 10 cm high, club-shaped, mainly solitary. Stipe yellowish, smooth, solid, gradually enlarging into a fertile cylindrical-ovoid head, at most 2.5 cm high and about half as wide, at first yellowish and smooth, becoming blackish and densely scabrous because of the emergence of immersed perithecia. Spores colorless, filiform, smooth, breaking up into secondary rodlike spores, 2.5–5 × 2 microns.
Edibility Of no interest.
Habitat On the hypogeal fruit-bodies of *Elaphomyces*.
Season Summer and autumn, like their host carpophores.
Note This species is best identified when dug up because it is attached to the false truffle indirectly by long, yellowish, cordlike strands, rather than, like *C. capitata*, directly.

351 TAPHRINA ALNI-INCANAE

Synonym *Taphrina amentorum.*
Etymology After its host, *Alnus incana,* gray alder.
Description This Ascomycete attacks various species of alder, causing a deformation on the floral cones with a consequent enlargement of the squamae and flowers, which leads to the formation of a sort of lamina that emerges from the cones. These galls reach a length of 1–3 cm, initially rose-brown, then curling over on themselves and, after a while, drying out and becoming light green quite tinged with red, and finally grayish-brown, easily detachable by breaking. The excrescences correspond to the same number of blooms which, as a result of the fungus attack, have proliferated by producing new tissue, on the surface of which the fungus produces, in a palisadelike formation, the bare asci. Because of the parasitic attack the entire inflorescence gradually assumes a squamose appearance, does not develop regularly, and rapidly dries out. Spores colorless, subglobose, smooth, 3–5 × 4–5 microns.
Edibility Of no interest.
Habitat Parasitic on alder inflorescences.
Season Summer.
Note A similar species, *T. deformans,* is more common, and causes peach "blister."

352 PILOBOLUS KLEINII

Etymology After the mycologist Klein.
Description Very small clubs called sporangiophores 0.5–0.7 cm high, transparent, erect, not ramified, enlarged in a small bladderlike formation toward top, at the tip of which there is a black, lenticular formation (the sporangium) containing the spores. At the base, immersed in the substratum, there is a small swelling.
Edibility Of no interest because of size.
Habitat In very dense groups on the droppings of various herbivores, principally cattle.
Season All year.
Note In sunlight a strong pressure is manifested in the sporangiophore, first by the emission of droplets of water and then by the explosion of the vesicle or bladder. Thrown some distance, in the direction of the light, the sporangium turns upside down in the air so that the lower, viscous part is at the front. This enables the sporangium to cling to the first object it touches. Normally the sporangia end up on blades of grass, enabling the fungus to spread to new substrata once they have passed through an animal's intestine.

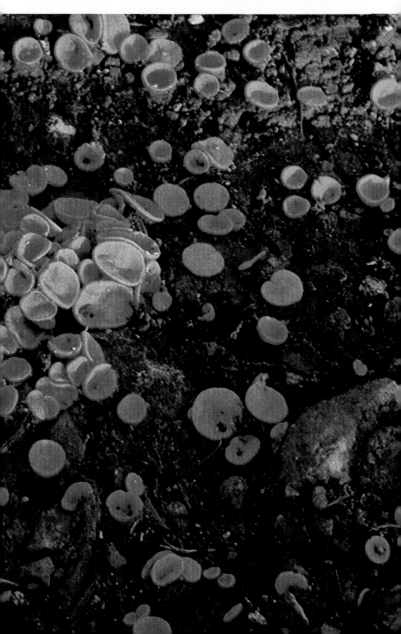

353 ANTHURUS ARCHER

Common name Archer's stinkhorn.
Etymology After the Irishman Archer.
Description Egg 4–6 × 3–4 cm, slightly buried in the ground, peridium pinkish white tinged with brown; when cut a gelatinous layer appears and, at the center, the pinkish white embryo of the carpophore is seen. Carpophore with 4–7 arms, cavernous, alveolate, fragile, red, 12 cm long, tapering pinkish white as soon as they have emerged. Gleba on the inner surface of the arms, dark greenish, mucilaginous, fetid. Spores grayish, elliptical, smooth, 5–7 × 2–2.5 microns.
Edibility Can be eaten at the egg stage, without the peridium and the gelatinous layer; odor of radishes.
Habitat In warm regions, on ground with plenty of rotting ligneous fragments, in acid soils, in broadleaf woods.
Season Summer.
Note Stinkhorns resemble puffballs when they first appear, although a cross-section of their egg reveals a heterogenous, usually multicolored context. As they develop they assume a variety of exotic shapes. Most are easily discovered by their odor before they are actually seen. They are disseminated by flies who feed on the slimy fetid spore mass.

354 ASERÖE RUBRA

Common name Starfish stinkhorn.
Etymology From Latin, "red," from arm color.
Description Egg up to 3 cm in diameter, whitish with pale yellow coloration; inside, a gelatinous stratum and, at center, the embryonic carpophore. Carpophore up to 6 cm high, when mature has a white then pinkish base emerging from volva, with 5–9 arms split in two after a few centimeters, up to 7.5 cm long and less than 0.5 cm wide at base, red. Gleba in central part of carpophore and slightly on base of arms, mucilaginous, fetid, dark olive-green. Spores within mucilage, grayish, elliptical, smooth, 4–5.5 × 1.5–2 microns.
Edibility Not known.
Habitat On humus-rich ground, with wood residue, in tropical and subtropical regions; in temperate regions it has been found in greenhouses.
Note Although not yet reported in North America it should be looked for in the Gulf Coast states as well as in conservatories throughout the country.

355 CLATHRUS RUBER

Synonym *Clathrus cancellatus.*
Etymology From Latin, "red."
Description Egg spheroidal, diameter up to 5 cm, white, with rhizomorphs at the base; in section, after the white, membranous, leathery peridium, a gelatinous stratum appears and, inside, the embryonic red carpophore with the greenish immature gleba. Carpophore globose, 5–10 cm high and 5 cm wide, red, with netlike lacunose arms delimiting polygonal holes. Gleba, when mature, is mucous, greenish black. Spores grayish, elliptical, smooth, 5–6 × 1.8–2 microns.
Edibility **Reportedly poisonous when eaten raw.**
Habitat Isolated or in small groups in woods in warm temperate regions; also in grasslands and greenhouses.
Season Spring to late summer.
Note *C. columnatus,* which is common at times in sandy soil in the South, has 2–5 stout curved arms joined together at their summit. *Pseudocolus schellenbergiae* grows in woodchip garden mulch in the East. It has 3 or 4 tapering arms attached at their tips, which rise out of a short stipe.
Caution Do not eat raw. See edibility.

356 DICTYOPHORA INDUSIATA

Etymology From Latin, "shift," referring to the veil.
Description Egg up to 4 cm in diameter, globose, ovoidal, white or grayish. Carpophore 15–20 × 2.5–3.5 cm, fusiform or cylindrical, barbed toward the top, white, porous, hollow, head ogival for a short time, then bell-shaped, yellowish under the gleba, white if stripped, with rugose surface, reticulate with apex perforated and delimited by a raised and distinct collar. Veil white, hanging almost to the ground, with wide polygonal chains formed by elliptical strands. Gleba olive-green, mucilaginous, not very fetid. Spores colorless, elliptical, smooth, 3.5–4.5 × 1.5–2 microns.
Edibility Reportedly eaten at the egg stage but not recommended.
Habitat Found in tropical forests.
Season During the rainy season.
Note *Dictyophora duplicata,* common in North America, rare in Europe, has a whitish head with distinctly raised reticulations and when fresh is covered by a fetid green slime. The veil is short with thinner chains and the peridium of the egg has a pinkish tinge.

357 MUTINUS CANINUS

Common name Dog stinkhorn.
Etymology From Latin, "canine."
Description Egg 2–4 × 1–2.5 cm, ovoid-elongated or pear-shaped, white or pale yellow with a mycelial crown at base, broken at apex into 2–3 lobes, from which the carpophore emerges. Carpophore 6–15 × 1 cm, cylindrical, hollow, first whitish then tinged with orange or faintly pinkish, slightly orange at apex when mature, fertile part beneath gleba bright red, sometimes perforated at top, ogival, 2 cm high, reticulate with deep, large cells. Gleba green and mucous, slightly fetid. Flesh cellular, fragile. Spores colorless, elliptical, smooth, 3–5 × 1.5–2 microns.
Edibility Of no interest.
Habitat In humus-rich ground rich in ligneous fragments, in damp parts of broadleaf woodland.
Season Summer to late autumn.
Note *M. elegans* has tapering carpophores that are up to 7 cm long in the red fertile part, and covered, like *M. caninus,* by a fetid green slime.

358 PHALLUS IMPUDICUS

Common name Stinkhorn.
Etymology From Latin, "shameless."
Description Egg 4–6 × 3–5 cm, spheroidal or elongated, white, with a white mycelial thread at the base, peridium white outside, tough, translucent inside, gelatinous. Carpophore up to 20 × 3 cm, tapering toward top, hollow, lacunose, porous, fragile, white, with a thick volva at the base, peridium, at top a conical cap, free at margin, covered with an alveolar reticulum, white with perforated disc at apex. Gleba olive-green, mucilaginous, fetid, positioned on cap from which it disappears in time, carried off by flies. Spores colorless, elliptical, smooth, 5–7 × 2–2.5 microns.
Edibility Reportedly eaten at the egg stage but **not recommended.**
Habitat In humus-rich ground, mainly beneath broadleaf trees, sometimes under conifers and in gardens.
Season Spring to late autumn.
Note *P. ravenelii* has a smooth to granular cap; *P. hadriani,* not reported in North America, grows in sandy soil, with a usually hypogeal egg. It is smaller and the peridium is conspicuously pale lilac-purple.

359 ASTRAEUS HYGROMETRICUS

Etymology From Greek, "which measures humidity," because of the behavior of the exoperidium.

Description Carpophore 2–4 cm, at first slightly hypogeal, globose, slightly viscous, star-shaped, open, epigeal, 6–10 cm wide. Exoperidium brown, opening into 6–10 or more arms, distending with humidity, closing inward when dry, made up of four strata, the innermost one tending to be deeply cracked. Endoperidium sessile, globose, flattened at the ends, diameter 2–3 cm, brownish, with irregular apical ostiole. Odor and flavor negligible. Spores chocolate-brown, round, warty, 9–11.5 microns.

Edibility Of no interest because of texture.

Habitat In warm, dry, temperate zones in clearings or open parts of woods.

Season Summer and autumn, but the carpophores remain more or less whole for about a year, with the exoperidium becoming increasingly thin.

360 GEASTRUM FORNICATUM

Etymology From Latin, "arched."

Description Carpophore open, 6–7 × 5–6 cm, light brown. Exoperidium opens in four triangular arms, the tips of which rest on the ends of four lobes of a structure consisting of a membranous mycelial stratum and the mass of mycelium which incorporates bits of wood and earth. Endoperidium subglobose, up to 2.5 cm in diameter, on a peduncle 2–3 mm high, open uppermost in a radiate peristome. Gleba powdery when mature. Odor and flavor negligible. Spores cocoa-brown, round, warty, 3.5–4 microns.

Edibility Inedible because of texture.

Habitat On sandy soils, in coniferous and broadleaf litter.

Season All year.

361 GEASTRUM PECTINATUM

Common name Earthstar.

Etymology From Latin, "with comblike teeth."

Description Carpophore first oval then exoperidium opens into 5–10 triangular lobes up to 6 cm wide, cracked in the middle, curved, white then ochreous outside and inside, with a fleshy stratum which breaks up and disappears. Endoperidium subglobose, 1–2.5 cm, brown or lead-colored, mealy, striate at base which extends into a thin peduncle, whitish to gray-brown, cylindrical, 0.6–0.8 × 0.2–0.3 cm, with raised peristome, deeply sulcate, frayed at apex. Gleba powdery and dark brown when mature. Spores brown, globose, evidently warty, 4–7.5 microns.

Edibility Of no interest because of texture.

Habitat In coniferous woodland and among bushes.

Season Summer and autumn, carpophores remaining unchanged for long periods.

Note Of the various species with the endoperidium supported by a peduncle and with a comblike peristome, *G. coronatum* has a dark brown peduncle and area of attachment, and like all the other species in the group except for *G. pectinatum*, it is not striate at the base of the endoperidium. *G. minimum*, which forms small carpophores up to 3 cm in stony grassland, has the peristome clearly delimited by a ringlike groove.

362 GEASTRUM FIMBRIATUM

Common name Earthstar.

Etymology From Latin, "fringed."

Description Carpophore up to 7 cm in diameter. Exoperidium opens into 6–9 triangular fringes, turned downward, chamois-colored, with an outer mycelial stratum, thin, membranous, separable, whitish. Endoperidium globose, sessile, same color as exoperidium, with fairly salient apical opening, edge fibrous. Gleba powdery when mature, cocoa-brown. Spores cocoa-brown, round, finely warty, 3–4 microns.

Edibility Inedible because of texture.

Habitat Under conifers.

Season Summer and autumn, although the carpophores remain on the ground, unchanged, for some time.

Note This fungus is easily identifiable because the endoperidium rests directly on the exoperidium without any peduncle, and because the peristome (the opening from which the spores emerge) is unevenly fringed.

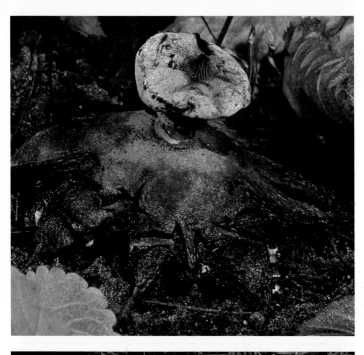

363 GEASTRUM TRIPLEX

Common name Earthstar.

Etymology From Latin, "triple," from the peridium.

Description Carpophore first subglobose, shaped like a tulip bulb. Exoperidium opening into 4–7 triangular lobes, up to 7 cm wide, fairly equal, divided in the middle, often cracked into areolae on the outside, olive-brown in the exoperidium, light brownish inside, with a thick fleshy stratum which breaks up and disappears except for the central part where a sort of cup forms at base of endoperidium, which is globose, 1.5–3.5 cm in diameter, light brown, sessile, membranous, with lighter apical ostiole, conical, fibrillose-fringed, with base not delimited. Gleba whitish at first, soon olive-ochre, powdery as soon as peridium opens. Spores brown, globose, warty, 4–5 microns.

Edibility Of no interest because of texture.

Habitat In woodland and grassland.

Season Summer and autumn.

Note The distinctive tulip-bulb shape (ovate with an acute apex) hallmarks the large carpophores of *G. melanocephalum* as well, which can reach a diameter of 22 cm when fully opened. This is a rare species, easily recognizable mainly by the fact that the blackish brown gleba is not enclosed in a membranous endoperidium, so that it becomes totally dispersed in the air.

---

364 MYRIOSTOMA COLIFORME

Etymology From Latin, "like a colander."

Description Carpophore 4–9 cm, at first hypogeal or semi-hypogeal, fairly globose, ochreous, often covered with fibrillose scales, brown or tinged reddish. Exoperidium opens up into a star, with 4–14 (usually 8) triangular arms, extending to 25 cm, feltlike outside, smooth inside, whitish then reddish brown, with transverse wrinkles. Endoperidium 1.5–7.5 cm, unevenly globose, flattened at the ends, surface shiny gray, tuberculate, then perforated with 30–50 ostioles, 1–2 mm wide, prominent; it is supported by a peduncle up to 5 cm high made up of about 12 small pillars which are separate or concrescent. Gleba white then cocoa-brown, divided into loculi, each of which opens into an ostiole. Spores reddish brown, globose, strongly warty, 4–6 microns.

Edibility Inedible because of texture.

Habitat Worldwide in sunny, dry woodlands, less frequent in open, sandy places.

Season Summer and autumn; carpophore remains can be found all year.

365 BOVISTA NIGRESCENS

Etymology From Latin, "blackening," from the color of the endoperidium.

Description Carpophore 1.5–5 cm, globose, with a mycelial tuft at the base. Exoperidium at first white, slightly pubescent, then with very faint areolations, darkening and finally becoming blackish when mature, detached from the endoperidium in the upper part. Endoperidium shiny, gray turning dark brown, opening at top into an irregular ostiole. Gleba white in immature specimens, cocoa-brown and powdery when mature. Odor and flavor slightly acrid in young specimens. Spores brown, globose, warty, 5–6 microns, pedicellate.

Edibility Mediocre while gleba remains white, then inedible.

Habitat Late spring to autumn.

Note *B. pila* is the North American equivalent of the European *B. nigrescens,* and differs in the field by the exoperidium, which darkens only to tan or pinkish.

366 BOVISTA PLUMBEA

Etymology From Latin, "leadlike," from the color of the endoperidium.

Description Carpophore 2–4 cm, subglobose, with no sterile base, attached to the soil by a small mycelial tuft. Exoperidium pure white, barely areolate, thin, fragile, soon separating from the endoperidium, at first yellowish, gray as soon as it opens, smooth, then blackish, opening to an apical ostiole, irregular. Gleba white, yellowish, powdery when mature, olive-brown with no basal sterile part. Flavor slightly acrid, in young specimens as well. Spores brown, oval, finely warty, 5–7 × 4–6 microns, pedicellate.

Edibility Mediocre while gleba remains white, then inedible.

Habitat In low grass in fields, on acid and basic ground, also in coastal dunes.

Season Spring to autumn.

Note The dry carpophore remains unchanged in fields and is carried off by wind; as it rolls around it is gradually destroyed by biological agents.

367 CALVATIA UTRIFORMIS

Synonyms *Lycoperdon caelatum; Calvatia caelata.*
Etymology From Latin, "like a wineskin."
Description Carpophore 7–15 cm or more in diameter, subspherical, globose, or pear-shaped. Exoperidium fragile, pure white, breaking up into polygonal areolae, pyramidal, which then disappear leaving the surface of the endoperidium areolate. The brown endoperidium, quite shiny when mature, opens unevenly, and up to half the carpophore disappears. Gleba white, then greenish ochre, finally dark olive-brown, at first compact then watery, finally powdery; subgleba with large cells, separated from the gleba by a pseudodiaphragm. Spores brownish, subglobose, smooth, 3.5–5 microns.
Edibility Very good while the gleba is immature (white).
Habitat Quite common in fields and meadows.
Season Summer and autumn.
Note When the spores have been released the rest of the carpophore—the subgleba with part of the endoperidium, a brown, papyruslike cup—remains in the field and is moved here and there by the wind. *C. craniformis* is very similar in the field and differs only microscopically; *C. cyathiformis* produces lilac spores.

368 CALVATIA GIGANTEA

Synonyms *Lycoperdon maximum; Langermannia gigantea.*
Common name Giant puffball.
Etymology From Latin, "gigantic."
Description Carpophore 10–65 cm in diameter, globose. Exoperidium white, thin, slightly mealy, pale ochreous when mature, minutely fragmented. Endoperidium thin, fragile, becoming fragmented and distintegrating. Gleba white, floccose, elastic when mature, olive-brown; subgleba almost absent. Thick, cordlike root. Spores olive-yellow, globose, smooth or barely warty, 3–5 microns.
Edibility Excellent when young, but see **caution.**
Habitat Worldwide, in meadows, fields, and gardens.
Season Summer and autumn.
Note This fungus can become very large indeed: weights of up to 56 pounds, diameters of 160 cm, and heights of 24 cm have been mentioned.
Caution Some people find this fungus indigestible.

369   CALVATIA EXCIPULIFORMIS

Synonym   *Lycoperdon saccatum.*
Etymology   From Latin, "vase-shaped."
Description   Carpophore higher than it is wide, of variable size up to 15 cm high, with a maximum diameter in the head of 10 cm and in the stipe of 6 cm. Stipe always well defined. Exoperidium white, cream-colored, or ochreous, granular or with small bristles converging at the apex. Endoperidium smooth, ochreous yellow, opening wide. Gleba initially white, when mature powdery and dark brown or cocoa-colored, globose; subgleba cellular, brown when mature. Spores cocoa-brown, globose, warty, 4–5.5 microns.
Edibility   Can be eaten when immature.
Habitat   In woods and fields with bushes.
Season   Summer and autumn.
Note   A fairly variable species, hence the numerous varieties that have been created on the basis of size, ornamentation, form, habitat, and color. The compact, feltlike, noncellular subgleba singles out two widespread species: *C. lilacina,* which turns from white to wine-red, and *C. candida,* white with gleba and subgleba not clearly defined.

---

370   LYCOPERDON PERLATUM

Synonym   *Lycoperdon gemmatum.*
Etymology   From Latin, "widespread."
Description   Carpophore 1.5–8 cm high, pear- or club-shaped, covered at top with 1–2 mm conical spines surrounded by warts or smaller spines which leave the endoperidium areolate after dropping off; the spines become increasingly smaller and thinner toward the bottom (but instead of the main spines there may be tufts of fine bristles very close together). The endoperidium opens into a circular hole at the top. Gleba at first white, ochre-brown when mature; subgleba well developed, with large cells. Spores yellowish brown, globose, smooth or barely warty, 3.5–4.5 microns.
Edibility   Mediocre when immature.
Habitat   Ubiquitous, on the ground, in woods and elsewhere, in lowland and upland areas.
Season   From late spring to late autumn.
Note   Emphasizing its polymorphism, numerous varieties have been created: *nigrescens,* with brown carpophores; *bonordeni,* the endoperidium of which is not areolated, with brown spines; *excoriatum,* with spines of just one sort with a pyramidal base; and *albidum,* with spines pyramidal or with a basal bulb, filiform at the base, with areolation.

371 LYCOPERDON UMBRINUM

Etymology From Latin, "dark-colored."
Description Carpophore 1–4 cm in diameter and up to 5 cm high, pear-shaped or subglobose. The exoperidium has scattered brown spines which reveal glimpses of the shiny, yellowish brown endoperidium; initially the spines are arranged in groups which converge at the apex and then usually separate. The endoperidium opens at the apex into a small hole. Gleba from white to cocoa-brown when mature; subgleba has large cells. Spores cocoa-brown, globose, warty, 4–5 microns.
Edibility Fairly mediocre.
Habitat Very widespread in the Northern Hemisphere, mainly in dry parts of coniferous and broadleaf woods.
Season Summer and autumn.
Note In dry, sunny grassland, especially in the Mediterranean countries, we find *L. decipiens,* with thin, soft, white spines, fairly convergent, which, as they drop off, reveal the shiny, white endoperidium. *L. mammiforme* is an impressive species covered with a white veil which breaks up into large caducous warts.

372 LYCOPERDON ECHINATUM

Etymology From Latin, "covered with spines."
Description Carpophore 2–7 cm in diameter, up to 5 cm high, pear-shaped or subglobose, brown, covered with spines joined together in threes, thus forming pyramidal warts up to 5 mm high, caducous, leaving behind them a continuous polygonal mosaic on the endoperidium, which opens out more or less circularly. Gleba white then ochre, soft then powdery and purple-brown; small subgleba with small, irregular cells. Spores chocolate-brown, globose, warty, 4–5 microns, pedicellate.
Edibility Edible when immature after removing exoperidium.
Habitat In woodland, usually on calcareous ground under beech.
Season Summer to autumn.
Note The variety *pulcherrimum* has a smooth endoperidium when the spines have dropped off. *L. candidum* is covered with warts formed by white spines, which fall in groups, with the endoperidium nonareolate.

373 LYCOPERDON PYRIFORME

Etymology From Latin, "pear-shaped."
Description Carpophore 1–5 cm in diameter, up to 8 cm high, pear-shaped or subglobose. Exoperidium formed by fine warts or broken up into plaques of varying shapes and sizes, rarely smooth, ochreous or hazel, darker at the apex. Endoperidium ochreous, tending to yellowish; on the whole the peridium is fragile when young, then becomes soft, opening at the apex where there is a papilla; at the base it has a white mycelial thread. Gleba white, then greenish yellow, olive, or brown; subgleba with small cells. Spores brownish yellow, globose, smooth, 3–4 microns.
Edibility Mediocre when young.
Habitat On rotten wood, often in large compact groups, rarely on ground which is rich in vegetable fragments.
Season Late spring to the first frosts.
Note *L. pyriforme* is an extremely variable species identifiable because: it prefers wood at its substratum; it has a mycelial thread at the base; it does not have distinct spines; and the small-celled subgleba is always white.

374 VASCELLUM PRATENSE

Synonyms *Lycoperdon hiemale; Vascellum depressum.*
Etymology From Latin, "of fields," from its habitat.
Description Carpophore 2–4 cm high, with equal diameter, subglobose or tubinate, sometimes with a well-defined stipe, typically white but more rarely yellowish brown or sometimes tinged orange-pink in young specimens, peridium covered with fragile, mealy ornamentation, either floccose elements or thin fairly connate spines. At maturity it opens at the apex into an ostiole at first small then gradually opening further, eventually affecting the whole fertile part. Gleba, when mature, is powdery and dark brown, separated from the sterile part quite distinctly by a diaphragm; subgleba cellular, reaching the median line of the carpophore, dark brown when mature. Immature flesh white. Spores light brown, subglobose or ovoid, spiny, 3–4.5 microns.
Edibility Mediocre while immature.
Habitat Widespread in temperate climates, often gregarious, in grassland, often with *Bovista plumbea* (**366**).
Season Spring to late autumn.
Note Although collected across northern North America, this mushroom is most commonly seen in lawns and grassy areas in the Pacific Northwest.

375 CALOSTOMA CINNABARINA

Etymology   From Greek, ''cinnabar.''

Description   Carpophore 5–9 cm high, 1.5–2.5 cm in diameter in the fertile part, globose, peridium orange-yellow, supported by a stipe, partly underground, venose-lacunose, whitish, gelatinous. A gelatinous peridium, like a transparent cap, covers the other two strata of the peridium; when mature it separates en bloc allowing the release of the spores from a crownlike peristome, orange-red in color. Gleba when mature is powdery and ochreous. Spores light ochre, elliptical, pitted, 10–15 × 8–10 microns.

Edibility   Of no interest because of texture.

Habitat   Partly buried in the ground (especially along road banks) in the Southeast, but extending up into New England.

Season   Late summer and autumn.

Note   Two other species with southern distribution, *C. lutescens* and *C. ravenelii*, differ from this mushroom by their lack of reddish color except at peristome; and they differ from one another in that the latter lacks a gelatinous exoperidium.

376 SCLERODERMA CITRINUM

Synonyms   *Scleroderma aurantium; Scleroderma vulgare.*

Common name   Common earth-ball.

Etymology   From Latin, ''lemon-yellow.''

Description   Carpophore diameter up to 12 cm, subglobose, peridium very thick, bright yellow, sometimes pale or creamwhite because the pigment is water-soluble, split into polygonal scales, coarse in texture, especially at the top, where it opens out irregularly. Gleba soon violet-black, blackish, finely marbled, pale greenish gray when mature, powdery. Odor and flavor strong and acrid. Spores brownish black, globose, spiny with faint reticulum that is chainlike, 7–15 microns.

Edibility   When eaten raw or even sometimes cooked it causes gastric disorders or acute indigestion; sometimes, cooked, very thin slices are used as pseudotruffles, but **not recommended**.

Habitat   Widespread, in woods, in acid or acidified soil.

Season   Summer and autumn.

Note   This mushroom is often seen parasitized by *Boletus parasiticus.* Several other species of *Scleroderma* prefer sandy soil, such as *S. macrorhizon,* with a deeply rooting, stalklike mycelial base, and *S. geaster,* which resembles a large earthstar when open.

377 TULOSTOMA BRUMALE

Synonym *Tylostoma mammosum*
Etymology From Latin, "wintery."
Description Carpophore 0.5–1 cm, spheroidal or sometimes depressed, whitish then pale yellow, outer peridium fragile, inner one smooth, thin, membranous, papyraceous, peristome small, slightly prominent, entire, brownish when mature. Stipe 2–5 × 0.2–0.3 cm, brownish, cylindrical, slightly tapering at top, where attached to peridium, base enlarged, bulblike because of the mycelial mass, smooth, or fairly fibrillose, brownish toward top. Gleba powdery when mature, ochreous. Spores pinkish yellow, globose, finely aculeate, 4–6 microns.
Edibility Of no interest because of texture.
Habitat Gregarious, with the stipe usually buried in moss in stony, calcareous areas, or in sandy soil.
Season Autumn through winter.
Note *Tulostoma* is a genus of stalked puffballs that develop underground and emerge at maturity. Often the stipe remains imbedded in the substrate. This and related Gasteromycetes with a puffball-like head and tough, fibrous stipe are to be expected in arid and semi-arid places and in dry, sandy soil.

378 HYDNANGIUM CARNEUM

Etymology From Latin, "flesh-colored."
Description Carpophore 1–3 cm in diameter, globose, surface pinkish to rose, then tending to yellow, ochreous in dry specimens, base sterile, like a small stipe, only evident in younger specimens. Gleba lacunose, fragile, marbled, with no particular odor or flavor. Spores pale yellow, globose, densely spiny, 10–18 microns.
Edibility Fair, although small.
Habitat Widespread in all warm temperate areas or in greenhouses beneath various broadleaf species, particularly eucalyptus.
Season Autumn to spring.
Note This species is associated essentially with eucalyptus trees, which are worldwide as a result of reforestation using these Australian trees. In warm climates the fungus can be found in winter and spring under the leaf litter.

379 GAUTIERIA MORCHELLAEFORMIS

Etymology From Latin, *"Morchella*-like (morellike)."
Description Carpophore 1–3 cm, sometimes 5 cm, globose, spheroidal, with no peridium in the adult state and with open hymenial cells, with a tiny stipe. Gleba ochreous rust-colored, consisting of broad cells 1.5–8 × 1–4 mm, powdery because of the spores. Flesh ochre-rust-colored, lacunose, fairly firm. Very strong odor of bouillon cubes and a sweetish flavor. Spores rust-colored, broadly elliptical, with 8–10 lengthwise ribs, 12–24 × 8–12 microns.
Edibility Eaten in Europe but **not recommended.**
Habitat In the Northern Hemisphere often found under broadleaf trees in acid soil, quite deeply buried and normally beneath leaves.
Season Summer and autumn.
Note This is one of a large number of fungi that grow underground. It is rare and not easily identified.
Caution Because very little is known about the edibility of similar species or of this one in North America, it cannot be recommended.

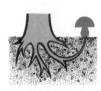

---

380 MELANOGASTER VARIEGATUS

Etymology From Latin, "speckled," because of the appearance of the gleba.
Description Carpophore 1.5–4 cm in diameter, spheroidal, elongated or humped, peridium at first yellow, ochre-rust-colored, downy, turning to blackish brown when touched, then yellowish brown, eventually dirty greenish brown. Gleba brown-violet, blackish, marbled with yellowish venations which emit a blackish mucous fluid. Odor at first of chicken liver then becoming more pleasant like liqueur chocolates, flavor sweetish, slightly sulfurous. Spores blackish brown, elliptical, smooth, 6–10 × 3.5–5.5 microns.
Edibility Eaten in Europe when young, but **not recommended.**
Habitat Hypogeal, symbiotic with broadleaf and conifer trees.
Season All year.
Note If exposed to the sun this fungus develops a very strong smell. The fluid which comes from the gleba originates from the hymenium which, when mature, is deliquescent, and the color is produced by the spores.
Caution Although eaten in Europe, it cannot be recommended in North America until more is known about this mushroom and its look-alikes and its local qualities are tested.

381 PISOLITHUS TINCTORIUS

Synonym *Pisolithus arhizus.*
Etymology From Latin, "used for dyeing."
Description Carpophore up to 30 cm high, at most 20 cm wide, subglobose or pear-shaped with a fairly long stipe, peridium dry, thin, smooth or faintly rugose or tuberculate, initially ochreous then dark brown, fragile and dehiscent when mature, beginning to open out at top into small polygonal scales. Base buried, sterile and solid, sturdy, 3–10 cm in diameter, often irregular, with, originally, a tuft of brown mycelial threads. Flesh at the top of the gleba formed by peridioles, at first compact, juicy, marbled in appearance, with violet-black peridioles with yellow or sulfur-yellow outline, starting to mature from top, becoming finally ochreous brown, and powdery. The sterile lower part has, initially, a marbled appearance, brown, white, or deep yellow, with minute areolas, ochreous brown when mature. Strong and pleasant mushroomy odor, flavor sweetish. Spores ochreous brown, globose, spiny, 9–12 microns.
Edibility Edible when immature, but unappealing.
Habitat Worldwide, in acid, sandy, thin soils; frequent in dry pinewoods, beneath juniper in dunes, in grassland, and in the remains of charcoal kilns.
Season Spring to late autumn.

382 RHIZOPOGON VULGARIS

Etymology From Latin, "common."
Description Carpophores 1.5–5 cm, globose, slightly lobate, at first white, then yellowish tinged with reddish, finally greenish brown, with mycelial threads, thin and ramified, adpressed, on the surface, peridium not detachable. Gleba soft, white at first, with narrow, mazelike cells, becoming greenish and finally olive-brown. Odor slightly acid and fruity, then acrid and penetrating, flavor initially sweet. Spores whitish, elliptical, smooth, 5–8 × 2–3 microns.
Edibility Can be eaten while immature, but difficult to identify to species in a genus where few species are known in terrms of their edibility.
Habitat Hypogeal beneath conifers, often gregarious.
Season Spring to late autumn.
Note *Rhizopogon* is a large genus of mushrooms that grow on or below the surface of the ground, are typically covered with rhizomorphs, and may at first be thought to be truffles, but truffles lack rhizomorphs and are very different microscopically. *Rhizopogon* is a Gasteromycete, producing its spores on basidia, whereas truffles are Ascomycetes, producing spores inside mostly globose asci.
Caution Difficult to identify. See edibility.

383   CYATHUS OLLA

Etymology   From Latin, a pot, from its shape.
Description   Carpophore 10–15 mm, with a maximum diameter of 8–15 mm, shaped like wide bell, gray or ochreous and faintly pubescent, then smooth outside, initially closed by a leathery diaphragm, whitish, fugacious, then open with an undulate margin erect or turned back, inner surface lead-gray or ochreous, smooth with peridioles (up to 10), circular in shape, compressed and like biconvex lenses, umbilicate, gray or blackish, shiny, up to 3–4 mm wide, attached with a white funiculus (cord) or peduncle. Spores white, broadly elliptical, smooth, 10–14 × 8–10 microns.
Edibility   Of no interest because of size.
Habitat   On thin soil, roots or rotting sticks, in flower pots.
Season   All year.
Note   This is one of the most widespread bird's nest fungi. Another quite frequent species is *C. stercoreus,* which normally grows on the droppings of large herbivores.

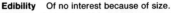

384   CYATHUS STRIATUS

Common name   Splash cup.
Etymology   From Latin, "with stripes."
Description   Carpophore 10–15 mm high with a maximum diameter of 8–10 mm, shaped like a truncated cone or bell, dark reddish brown, covered externally with evident hairs, initially with the upper part closed by whitish diaphragm, fugacious, then open, opening into a shiny, lead-colored cavity, sulcate lengthwise, with, on the inside, whitish-gray peridioles, circular, compressed, umbilicate and about 2 mm in diameter, attached by a white, elastic funiculus. Spores white, elliptical, smooth, 18–22 × 8–12 microns.
Edibility   Of no interest because of size.
Habitat   Isolated or in large groups, on dead wood or on the ground.
Season   All year.
Note   The spores are contained inside the peridioles, which are expelled by a drop of rain which manages to enter the carpophore and splash them out. In the air the funiculus, when completely distended, acts like a whip, enabling the peridiole to become attached to blades of grass.

385 CRUCIBULUM LAEVE

Synonyms *Crucibulum levis; Crucibulum vulgare.*
Description Carpophore 5–8 mm high, with a maximum diameter of 6 mm, at first subglobose then campanulate or cylindrical, grayish, ochreous brown, or ochre, initially closed by a fugacious diaphragm, then opening wide, finely downy on the outside then glabrous, whitish, smooth on the inside with whitish, circular, biconvex peridioles, 1.5–2 mm wide, attached to a papilla by a funiculus or peduncle. Spores white, elliptical, smooth, 5–10 × 4–6 microns.
Edibility Of no interest because of size.
Habitat Gregarious on sticks, wood, stems of herbaceous plants.
Season Spring to late autumn.
Note The carpophore consists of two distinct layers which can be separated. The peridioles develop and for some time remain in a gelatinous liquid. This is one of the commonest bird's nest fungi.

386 SPHAEROBOLUS STELLATUS

Etymology From Latin, "star-shaped."
Description Carpophores 2 mm wide, whitish or pale yellow, downy then smooth, globose then oval and opening to form a star, splitting into 6–8 triangular lobes, allowing the expulsion of the gleba in a gelatinous spheroidal mass, at first whitish and transparent then brown, broadly elliptical. Outer surface covered by a sparse mycelial veil. Spores white, elliptical, smooth, 8–11 × 4–6 microns.
Edibility Of no interest because of size.
Habitat Typically gregarious on rotting wood, sticks, leaves, or sawdust.
Season All year.
Note The multistratified peridium which splits open like a star allows for the active expulsion of the spores gathered in a gelatinous globule. On the basis of size and habitat various similar species have been described which may be considered as varieties of *S. stellatus: giganteus,* reaching 4 mm in diameter; *solen,* with cylindrical carpophores; *stercorarius,* globose, completely buried in droppings of herbivores; *muscosus,* covered with an abundant mycelium; *brasiliensis,* like the above with larger spores, 10–12 × 6–7 microns.

387 ELAPHOMYCES GRANULATUS

Etymology From Latin, "granular," because of the appearance of the peridium.

Description Carpophores up to 4 cm in diameter, globose or ovoid, outer surface of the peridium ochreous, covered with small, sometimes pyramidal warts. When cut it reveals a thick peridial "bark" formed by two layers, the outer one thin and yellowish, containing the warts, the inner one whitish and thicker. Gleba violet-black, at first divided by whitish veins of sterile tissue, then powdery. Spores blackish brown, spherical, covered with low warts, 24–32 microns.

Edibility Undetermined.

Habitat Semihypogeal on the ground or in moss, in acid soils, beneath pine, also under broadleaf species.

Season Fall through early spring.

Note This Ascomycete is the commonest of the hypogeal or semihypogeal fungi which resemble truffles in appearance, but which are not reliably known to be edible. This mushroom is best found by finding one of two conspicuous parasites on it: *Cordyceps capitata* (**349**), or *C. ophioglossoides* (**350**).

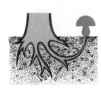

---

388 TUBER AESTIVUM

Common name Summer truffle.

Etymology From Latin, "of summer."

Description Carpophores normally 2–10 cm, sometimes larger, globose or irregularly lobate or deformed, with or without a small basal cavity, brownish black, with a thin peridium and large pyramidal warts with 5–7 faces up to 12 mm wide and 3 mm high, thinly striate concentrically and grooved lengthwise. Gleba at first whitish, light ochreous brown, then brownish, marbled by numerous white veins, ramified and anastomosed, at first firm then rather soft. Odor increasingly strong and aromatic with age, flavor distinctive and pleasant. Spores brownish yellow, subglobose, alveolate-reticulate, with spines in the nodal points of the reticulum, 18–41 × 14–32 microns.

Edibility Very good, even though tougher than the most prized truffles.

Habitat Hypogeal in calcareous ground, under broadleaf trees, prefers oak, also under pine and juniper.

Season All year, less frequent in spring, plentiful in summer.

Note Widespread in Europe and North Africa.

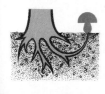

389 TUBER MELANOSPORUM

Common names French truffle; Perigord truffle.
Etymology From Greek, "black-spored."
Description Carpophores variable in size, normally 2–7 cm, globose or irregular, peridium at first reddish then dark black, warty with quite small pyramidal warts with 4–6 faces. Gleba tender, faintly watery, first grayish then violet, blackish when mature, marbled with whitish veins, quite ramified. Very strong odor of garlic and acetylene, flavor complicated to describe. Spores blackish, elliptical, spiny, 30–50 × 20–30 microns.
Edibility Excellent, raw or cooked.
Habitat Hypogeal in red calcareous ground containing iron salts, symbiotic with broadleaf trees (especially oak) in southern Europe and in the Perigord region of France.
Season The carpophores start to form in late summer, but mature between November and March.
Note Unfortunately not known to occur in North America. This famous truffle is one of the high points of gastronomy. Related species are *T. aestivum* (**388**), *T. brumale*, *T. moschatum*, *T. macrosporum*, and *T. mesentericum*, all black truffles, less sought after although all good to eat; they have a wider distribution than *T. melanosporum* itself.

---

390 TUBER EXCAVATUM

Etymology From Latin, "hollow."
Description Carpophore 2–5 cm, globose, often lobate, perforated at base with an internal cavity; peridium yellow, ochreous, olive, reddish orange, or brown, with tiny papillae or almost smooth, hard, ligneous. Gleba at first ochreous, reddish or orange, brownish when mature, with few whitish or yellowish veins, tough, horny. Strong garlicky odor, pleasant flavor. Spores brownish ochre, elliptical, reticulate-alveolate, 22–55 × 16–40 microns.
Edibility Although hard it can be eaten grated.
Habitat On dry, sunny calcareous ground, usually under broadleaf species, especially oak.
Season All year, more frequent in summer and autumn.
Note *T. fulgens* and *T. monticellianum* are very similar species, the former rust-colored, orange-yellow, or saffron, with a granulose peridium, and the gleba orange pink then blackish purple; the latter is softer, with a smooth, dark olive-brown peridium, gleba blackish brown with thin venations. *T. rufum*, which is quite variable in color, externally resembles the above-described species, but is solid. There is a hint of hollowness at the base of *T. nitidum*.

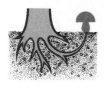

391 TUBER MAGNATUM

Common name White truffle.
Etymology From Latin, "lordly."
Description Carpophore variable, normally 2–8 cm, globose, regular or irregularly lobate, peridium ochreous, pale ochreous yellow, tinged greenish, smooth or faintly papillate, thin and adherent. Gleba white, yellowish, then yellowish gray, reddish brown, reddish gray, soft, tender, marbled with thin white veins, ramified and anastomosed. Penetrating odor of garlic, flavor complex, strong, and pleasant. Spores brownish, globose or broadly elliptical, with large reticulum, 35–50 × 32–42 microns.
Edibility Excellent, cooked or raw.
Habitat Hypogeal in crumbly calcareous ground, beneath oak, poplar, willow, and lime.
Season Summer through winter.
Note This highly prized hypogeal fungus has a very limited distribution: in central-northern Italy, in the Swiss canton of Ticino, and in France's Rhone Valley.

392 TERFEZIA ARENARIA

Synonym *Terfezia leonis.*
Etymology From Latin, "pertaining to sand."
Description Carpophore 6–12 cm in diameter, globose or pear-shaped, with a small short conical stipe, peridium smooth, whitish then yellowing, then darkening after turning reddish, quite spotted, often cracked. Gleba white, yellowish, rose-colored, or red, then grayish brown, marbled with white. Odor faint, flavor pleasant, sweetish. Spores at first whitish, eventually ochreous, globose, very warty, 19–26 microns.
Edibility Good, very popular in desert regions of North Africa and the Middle East.
Habitat In sandy ground, grassland, or open places in oak woods, pinewoods and stands of eucalyptus, symbiotic with cistus and rockrose.
Season Winter and spring.
Note This is the famous truffle of classical antiquity (the Greek *hydnon* and the Roman *tuber*). It is still highly prized in Islamic countries, where it is called *terfaz,* which term embraces various other species of hypogeal fungi. The other principal truffles which grow in a sandy habitat are members of the genera *Lespiaultinia, Delastria,* and *Tirmania.*

393 DALDINIA CONCENTRICA

Etymology From Latin, "concentric."

Description Stromata up to 5 cm in diameter, semispherical, sometimes incurved at the base, at first reddish brown, but soon turning black, smooth, shiny, dotted with the pores of the perithecial ostioles which form a circular layer beneath the crust. Flesh hard, fibrous, purple-brown, with darker concentric zones. Spores black, elliptical, fusiform, smooth, 12–17 × 6–9 microns.

Edibility Inedible because of texture.

Habitat On dead broadleaf wood, especially ash.

Season All year.

Note In the first stage of growth reddish asexual spores called conidia are produced which initially color the outer surface of the stromata; when mature these are replaced by black ascospores.

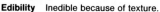

394 HYPOXYLON FRAGIFORME

Etymology From Latin, "strawberrylike," because of the appearance.

Description Stromata about 1 cm in diameter, semispherical, salmon-pink, when mature brick-red, darkening with age, surface papillate corresponding to the upper part of the perithecia. Flesh hard, black, with a layer of small flask-shaped perithecia just beneath the outer surface. No odor or flavor. Spores dark brown, blackish, almost fusiform, smooth, 11–15 × 5–7 microns.

Edibility Of no interest because of texture.

Habitat Normally gregarious on rotting beech branches.

Season All year, but grows in summer and autumn.

Note All the species of this genus grow in wood residue; many form crusty stromata; hemispherical stromata are formed by *H. fraxinophilum* (brownish gray on ash branches) and *H. fuscum* (on hazel or alder branches).

395 LYCOGALA EPIDENDRUM

Common name   Toothpaste slime mold.
Etymology   From Greek, "on trees," from its habitat.
Description   Carpophores up to 2 cm in diameter, sessile, globose, at first orange-red, soft and viscous, with a liquid interior, then brown with a peridium which becomes membranous, dry, with the inside turning into a brownish powdery mass. Spores grayish brown, globose, reticulate, 4–6 microns.
Edibility   Of no interest because of texture.

Habitat   On rotting coniferous and broadleaf wood and stumps.
Season   All year, but more common in summer and autumn.
Note   This belongs to the class *Myxomycetes*, organisms difficult to place in the systematics of the division Fungi. Their cycle includes a free, single-celled, creeping, amoeboid stage; then, because of the hormones produced by some of the "amoebas," they join together and differentiate themselves, developing a membrane and eventually forming fruit-bodies called peridium or sporangium, which contain tough-walled spores that are extremely resistant and long-lived (germination has been observed after 61 years of conservation). The characteristics of these organisms place them somewhere between animals (Protozoans) and the true fungi (Eumycetes).

396 SARCOSPHAERA CRASSA

Synonym   *Sarcosphaera coronaria.*
Etymology   From Latin, "thick."
Description   Carpophore spheroidal, opening into a circular hole in the upper part, hypogeal, like an empty ball, then stellate, revealing its violet then violet-brown hymenium. The mature carpophores, when completely expanded, are up to 20 cm wide. Outer surface develops from a viscous stratum which starts splitting from the orifice where it appears like a cobwebby veil that is detachable while moist, then adheres closely to outer part of carpophore which is dirty white, finely downy, sometimes marked with yellow especially toward base, normally with earth residue. Flesh white, fragile, waxlike. No particular odor or flavor. Spores white, broadly elliptical, smooth 13–14 × 7–8 microns.

Edibility   Edible cooked, but indigestible for some; **poisonous raw.**
Habitat   Common in pinewoods, on calcareous ground, also beneath various broadleaf species such as beech and oak, also on clayey ground.
Season   Spring.
Note   *Peziza ammophila* forms carpophores half-immersed in sand. They open out like stars and reveal a brown hymenium.
Caution   Poisonous when raw.

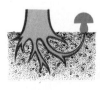

397 BULGARIA INQUINANS

Etymology From Latin, "dirtying," because if one touches the mature carpophores spores stain the fingers blackish.

Description Carpophore 1–5 cm wide, up to 3 cm high, initially almost ovoid, when mature like an upturned cone, with the upper fertile part gradually expanding. The hymenium is at first concave then flat, smooth, shiny and black; outer surface violet-brown, with granules which, as they become detached, leave behind whitish areolae. Flesh thick, gelatinous, elastic, marbled ochreous brown. No particular smell or flavor. Spores soot-brown, kidney-shaped with pointed tips, smooth, 10–13 × 6–7.5 microns.

Edibility Rather mediocre.

Habitat On rotting broadleaf wood, mainly oak.

Season Late summer and autumn.

Note Half the spores are colorless because they fail to mature.

398 PHAEOHELOTIUM SUBCARNEUM

Etymology From Latin, "almost flesh-colored."

Description Carpophores 0.1–0.2 cm, cup-shaped or flat, with a small stipe, smooth, soft, entirely pale purple-pink. Spores brown, elliptical, smooth, 10–12 × 2.5–4 microns.

Edibility Of no interest because of size.

Habitat In groups on wood stripped of bark.

Season Late summer and autumn.

Note Wood or vegetable detritus offers an important substratum for a great many Ascomycetes, with small cup-shaped carpophores, almost gelatinous, elastic, supported by a peduncle or sessile. A related species, *P. flavum*, is yellow, and its white-spored look-alike is the ubiquitous *Bisporella citrina* (**399**). Numerous other small gelatinous-elastic cup-shaped Ascomycetes occur on wood; they are classified in various genera and their identification requires microscopic analysis. An equally large grouping occurs on leaves and stems, especially on fallen, rotting herbaceous material.

399 BISPORELLA CITRINA

Synonyms *Calycella citrina; Helotium citrinum.*
Etymology From Latin, "lemonlike," because of the color.
Description Carpophore up to 3 mm in diameter, deep yellow, slightly paler on the outside, tending to orange-yellow in dry weather, completely glabrous, with a fairly long peduncle; surface fertile, initially barely concave, then almost flat. Spores colorless, elliptical, sometimes with a single septum, smooth, 9–14 × 3–5 microns.
Edibility Of no interest because of size.
Habitat Gregarious on dead broadleaf wood, often erupting from the bark.
Season Spring to autumn.
Note This fungus is similar in appearance to species belonging to the genus *Hymenoscyphus,* which grow on herbaceous substrata, not on wood.

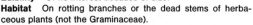

400 DASYSCYPHUS SULFUREUS

Synonym *Lachnella sulfurea.*
Etymology From Latin, "sulfur-colored."
Description Carpophore 1–3 mm, sessile, hemispherical then flat, sometimes slightly undulate, hymenial surface white or whitish, outside covered with lemon- or sulfur-yellow hairs, tending to darken all over with age. Spores colorless, rodlike or very fusiform, often slightly curved, smooth, 12–20 × 2–3 microns.
Edibility Of no interest because of size.
Habitat On rotting branches or the dead stems of herbaceous plants (not the Graminaceae).
Season Summer and spring.
Note Among the many species which form small cup-shaped carpophores covered with hairs, easily identifiable are: *D. virgineus,* with peduncle, completely white and very common on vegetable fragments; *D. bicolor,* with a very short stipe yellow inside and white outside; *D. auroreinus,* pedunculate, orange-pink, growing on stems of herbaceous plants; *D. flavovirens,* with a bright yellow hymenial surface, outside olive-brown, growing on larch branches; *D. fagicolus* and *D. hippocastani,* the former brown and growing on beech cupules, the latter violet and growing on chestnut husks.

401 HYMENOSCYPHUS FRUCTIGENUS

Synonym *Helotium fructigenum.*

Etymology From Latin, "growing on fruit."

Description Carpophore up to 4 mm in diameter, entirely light yellow, fertile surface flat, peduncle and outer surface glabrous or finely pubescent with adpressed filaments. Spores colorless, often septate in the middle, elliptical, smooth, 13–25 × 3–5 microns.

Edibility Of no interest because of size.

Habitat On the husks of rotting fruit of beech, hazel, hickory, and oak; also reported on chestnuts and cherry stones in Europe.

Season All year, but generally found late summer and autumn.

Note There are many similar species, and many of them are easily identifiable by their habitat: *H. scutula,* growing on the stems of composites and mints; *H. repandus,* on the dead stems of thistle rush, and iris, and on the husks of horse chestnuts; *H. caudatus,* typically found on the rotting leaves of various broadleaf species; *H. egenulus,* on dead stems of sheep sorrel; and *H. robustior,* on the dead stalks of marsh plants.

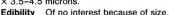

402 CHLOROSPLENIUM AERUGINOSUM

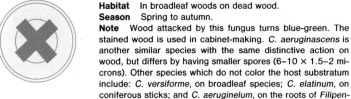

Common name Greenstain.

Etymology From Latin, "copper-green," from its color.

Description Carpophore up to 6 mm in diameter, bluish green, hymenium becoming paler with age, often yellowish as well, edge irregularly undulate, stipe fairly short and glabrous. Spores colorless or faintly pale green, elliptical, smooth, 9–15 × 3.5–4.5 microns.

Edibility Of no interest because of size.

Habitat In broadleaf woods on dead wood.

Season Spring to autumn.

Note Wood attacked by this fungus turns blue-green. The stained wood is used in cabinet-making. *C. aeruginascens* is another similar species with the same distinctive action on wood, but differs by having smaller spores (6–10 × 1.5–2 microns). Other species which do not color the host substratum include: *C. versiforme,* on broadleaf species; *C. elatinum,* on coniferous sticks; and *C. aeruginelum,* on the roots of *Filipendula.* All are green.

403 ASCOCORYNE SARCOIDES

Synonym *Coryne sarcoides.*
Etymology From Greek, "fleshy."
Description Carpophore 1–1.5 cm high, sessile or with a short stipe, violet, pale lilac, or pale purple, cup-shaped, slightly concave at first, then open, undulate, lower part clearly venose and sometimes pruinous. Flesh fleshy-gelatinous, violet. No particular odor or flavor. Spores white, fusiform, slightly curved, smooth, septate when very mature, 10–20 × 4–6 microns.
Edibility Of no interest because of size.
Habitat On dead broadleaf wood.
Season Autumn and winter.
Note Because of its appearance this species may be taken for a jelly fungus, but it is in fact an Ascomycete because its spores are produced inside saclike structures (asci) rather than on the outside of clubs (basidia). This and other jellylike Ascomycetes lack the horny-tough texture of most true jelly fungi.

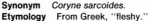

404 DISCIOTIS VENOSA

Etymology From Latin, "veined," from its appearance.
Description Carpophore 4–20 cm in diameter, initially subglobose with the margin incurved, then open and undulate, flat, fairly dark brown, venose uppermost, whitish and tomentose below. Stipe short, thick and sulcate. Flesh very brittle, and readily breaking. Slight odor of chlorine, virtually no flavor, although slightly bitter and astringent. Spores whitish, elliptical, smooth with caducous granulations at the tips, 21–25 × 12–15 microns.
Edibility Good when thoroughly cooked; **poisonous raw.**
Habitat On the edge of woodland and in open places.
Season Spring and summer.
Note Other cup fungi, such as *Discina perlata* and *Peziza repanda,* may resemble this fungus when mature with the carpophore completely expanded.
Caution Like many cup fungi, this mushroom is poisonous if eaten raw or undercooked.

405 HUMARIA HEMISPHAERICA

Etymology From latin, "hemispherical."

Description Carpophore up to 10–20 mm wide, at first sub-globose, then cup-shaped and hemispherical, hymenial surface smooth, grayish white, ochreous and covered with brown hairs, fasciculate, pointed externally, edge with hairs up to 1 mm long. Flesh thin, waxlike. No odor or flavor. Spores whitish, elliptical, warty when mature, 21–25 × 11–13 microns.

Edibility Of no interest because of size.

Habitat In damp, humus-rich ground, in woodland, sometimes also on fairly decomposed wood.

Season Summer and autumn.

Note Two similar species that are saucer-shaped instead of cup-shaped are *Trichophaea woolhopeia*, which grows in burnt places; and *T. hemispheroides*, which has the same habitat but seems to be associated with mosses of the genus *Funaria*.

406 PAXINA ACETABULUM

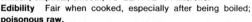

Synonyms *Acetabula vulgaris; Helvella acetabulum.*

Description Carpophore up to 12 cm high and 6 cm wide, cup-shaped. Hymenium violet-brown, outer surface lighter, edge undulate and turned inward, outer surface finely furfuraceous, pustular toward edge, becoming darker brown with age. Stipe up to 6 cm high, with ramified ribs that envelop the cup in the upper part, whitish at base, darkening from bottom to top where it is the same color as the cup. Flesh waxlike, quite elastic in the stipe, white. No particular odor or flavor. Spores whitish, elliptical, smooth, 18–22 × 12–14 microns.

Edibility Fair when cooked, especially after being boiled; **poisonous raw.**

Habitat In open places, on the edge of woodland, in sandy soils; solitary or gregarious, worldwide.

Season Spring, sometimes also late summer.

Note Mushrooms variously called *Acetabula* and *Paxina* are indistinguishable microscopically from *Helvella;* their fruit-bodies differ by being cup-shaped rather than saddle-shaped.

Caution Poisonous when raw.

407 PAXINA BARLAE

Synonym *Acetabula barlae.*
Etymology After the French mycologist Barla.
Description Carpophore 3–5 × 3–5 cm, cup-shaped, hymenial surface slate-gray or blackish violet, externally soot-colored, dark gray, blackish, sometimes olivaceous, finely furfuraceous. Stipe short, quite stout, white, sulcate with lengthwise ribs which become extended, becoming cup-colored, halfway up ramifying. Flesh whitish, waxlike. No odor or flavor. Spores white, elliptical, smooth, 20–22 × 11–12 microns.
Edibility Good.
Habitat Isolated or in groups beneath or near pines on calcareous ground.
Season Spring.
Note Much more common is *P. leucomelas,* which has a whitish outside and stipe and a dark grayish brown hymenium. *P. sulcata* has a blackish brown hymenium and grayish to brown outside and stipe.

---

408 TARZETTA CATINUS

Synonyms *Geopyxis catinus; Pustularia catinus.*
Etymology From Latin, "small bowl," from its shape.
Description Carpophore 1–5 cm wide, cupulate, permanently goblet-shaped, often with short, slightly wrinkled and spongy stipe, cream-colored internally, exterior yellowish-ochre and finely granular, with dentate, serrated margin. Flesh thin, fragile. No particular odor or flavor. Spores white, elliptical, smooth, 20–26 × 11–15 microns.
Edibility Mediocre.
Habitat Solitary or in groups more or less embedded in the humus in coniferous woodland.
Season Spring.
Note In the early stages of development the cup orifice is covered by a weblike cortina. A similar species but grayish tan and typically not over 2 cm wide is *T. cupularis;* it is also found in spring in coniferous woods, often in moss. *Geopyxis vulcanalis* is another very small cup with a crenate margin. It is yellowish but essentially differs microscopically from species of *Tarzetta:* it lacks oil drops in its spores, whereas *Tarzetta* has two large oil drops.

409 ALEURIA AURANTIA

Synonym *Peziza aurantia.*

Common name Orange peel peziza.

Etymology From Latin, "orangelike," from its color.

Description Carpophore up to 10 cm in diameter, sessile, cup-shaped, margin undulate, when mature flat, irregular, often deformed by the mutual compression of adjacent carpophores. Hymenial surface bright orange-red above, finely pubescent and whitish below. Flesh thin, white, fragile. No clear odor or flavor. Spores white, elliptical, reticulate, 17–24 × 9–11 microns.

Edibility Good cooked.

Habitat Gregarious, on bare ground in woodland, on the side of roads, in grassy places.

Season Spring, and autumn through winter.

Note The red pigment which colors this best-known of the cup fungi is quite common among the Discomycetes and is produced by pigments similar to the carotenoids found in the plant kingdom. Sunlight is necessary to produce these pigments.

---

410 OTIDEA ONOTICA

Etymology From Greek, "like an ass's ear."

Description Carpophore up to 10 cm high and 6 cm wide, often gregarious, variable in form, sometimes because of mutual compression, or shaped like the pavilion of the ear of an herbivore, with a short stipe; inner fertile part ochreous tinged with pink, becoming deeper in color as it dries out; outer sterile surface finely furfuraceous, at first almost the same color as inner part, then clearly ochreous. Stipe white at the base. Flesh thin, white. No particular odor or flavor. Spores white, broadly elliptical, smooth, 10–13 × 5–6 microns.

Edibility Fair.

Habitat On the ground, often in tight groups, especially in broadleaf woods, particularly oak.

Season Summer and autumn.

Note Other ear-shaped cup fungi include yellow-brown *O. leporina;* yellowish red-brown *O. smithii;* and the typically clustered *O. alutacea,* with a yellow-brown exterior and grayish brown interior. None of these has a pink to rosy inner ear, and all grow in coniferous woods.

411   SCUTELLINIA SCUTELLATA

Common name   Eyelash peziza.

Etymology   From Latin, "bowl-shaped" or "flat."

Description   Carpophore 2–10 mm, sessile, first slightly concave then flat, vermilion on the fertile upper surface, paler on lower surface, margin, with long stiff, dark hairs. Spores white, elliptical, slightly granulose outside, 18–19 × 10–12 microns.

Edibility   Of no interest because of size.

Habitat   Gregarious on decomposed wood or damp ground.

Season   Spring to late autumn.

Note   Although small, this is a fairly conspicuous fungus both because of its bright color and because it grows in groups. A very similar fungus called *S. trechispora*, with globose, warty spores, also grows on damp ground, often in moss. Other small Discomycetes with an outer ciliate surface belong to the genera *Sphaerosporella*, with a brown fertile surface; and *Cheilymenia* and *Neottiella*, with pale yellow or transparent hairs and the hymenium respectively yellow and red. The sterile surface of the sometimes terrestrial genus *Melastiza* is covered with short cilia; the same feature applies to the various species of *Anthracobia*, found in burnt places.

412   PEZIZA VESICULOSA

Etymology   From Latin, "with vesicles," because of the crenulate edge and the outer surface.

Description   Carpophore 3–8 cm, initially subglobose, barely open at the top, then hemispherical with incurved margin, clearly crenulate and furfuraceous, then distended and cracked; light yellowish ochre inside, whitish and at times tinged with reddish brown; furfuraceous and downy outside, sessile or with a short stipe. Flesh light ochreous brown, juicy, thick, fragile. No odor or flavor. Spores white, elliptical, smooth, 20–25 × 12–14 microns.

Edibility   Fair.

Habitat   Solitary or tufted, on the droppings of large herbivores or on the ground.

Season   All year, especially in late winter and spring.

Note   The carpophores vary somewhat both microscopically and microchemically depending on their particular habitat.

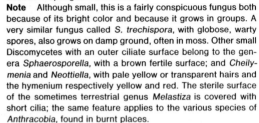

413 RHIZINA UNDULATA

Etymology From Latin, "wavy," from the form of the carpophore.

Description Carpophore up to 10 cm wide, flat or convex, irregularly lobate, fertile surface undulate, dark brown, sometimes blackish, with the margin paler; lower surface ochreous, with numerous cylindrical and ramified structures like small roots, 1–2 mm thick, which attach the carpophore to the ground. Flesh reddish brown, slightly fibrous, then rather leathery. Spores whitish, fusiform, pointed at the tips, minutely verrucose, 22–40 × 8–11 microns.

Edibility Of no interest because of texture.

Habitat Gregarious on the litter in sunny coniferous woods, often in burnt areas.

Season Summer and autumn.

414 RUTSTROEMIA ECHINOPHILA

Etymology From Greek, "spine-loving," because of its habitat (on chestnut husks).

Description Carpophore 2–12 mm in diameter, cup-shaped, flat or slightly convex, reddish brown, sometimes tinged pale purple, initially with the edge finely furfuraceous and denticulate, then entirely smooth. Stipe slender, 2–12 mm long, when young covered with fibrils, slightly paler than the hymenial surface, base blackish brown, emerging from blackened areas inside the rotting husk. Spores whitish, elliptical, slightly curved, smooth, 18–20 × 5–6 microns, with one or two septi, and sometimes a secondary spore.

Edibility Of no interest because of size.

Habitat Emerging from the inside of the husks of sweet chestnut, sometimes on the acorns of oaks.

Season Late summer and early autumn.

Note Other brown species of *Rutstroemia* have different habitats: *R. sydowiana,* on the venations and stalks of oak leaves; *R. conformata,* on alder leaves; *R. fruticeti,* on blackberry bushes and brambles; *R. firma,* on oak branches; and *R. rhenana,* on apple and pear branches. The greenish *R. luteovirescens* grows on maple and linden leaves.

415 SARCOSCYPHA COCCINEA

Common name Scarlet cup.
Etymology From Latin, "vermilion," because of the color.
Description Carpophore 1–5 cm in diameter, cup-shaped, fairly open, scarlet or orange inside, pinkish and downy outside, margin slightly incurved. Stipe absent or up to 2 cm high, villose at the base. Flesh waxlike and elastic. No particular odor or flavor. Spores whitish, elliptical, smooth, 24–40 × 10–12 microns.
Edibility Mediocre.
Habitat On rotting branches, buried in humus, or by watercourses.
Season Winter or early spring.
Note *S. occidentalis* has a smaller fruit-body and longer stipe. Both are among the first mushrooms to be seen in the early spring.

416 SEPULTARIA SUMNERIANA

Etymology After the mycologist Sumner.
Description Carpophores 3–7 cm in diameter, at first subspheroidal, barely open uppermost, hypogeal, then emerging, opening on the surface of the ground in irregular lobes, whitish and slightly ochreous inside, covered on the outside with down consisting of very long, septate filaments which incorporate loam. Flesh compact, two-layered, the hymenial one translucent, the outer one (beneath the hairy covering) white. No odor or flavor. Spores white, fusiform, smooth, 30–35 × 14–15 microns.
Edibility Edible after removing outer layer.
Habitat Hypogeal in lawns under cedar and yew.
Season Late winter and spring.
Note This is common in Europe, specifically associated with cedars. There are other, somewhat rare species of *Sepultaria*: *S. foliacea*, with the hymenial surface light greenish yellow, growing in moss; *S. tenuis*, a grayish white inside, opening out in roundish laminae; *S. arenosa*, initially ovoid, with triangular laminae; *S. arenicola*, brownish inside. The genus *Sepultaria* represents a transitional form of the Discomycetes from the epigeal to the hypogeal environment.

417 MACROSCYPHUS MACROPUS

Synonym *Helvella macropus.*
Etymology From Greek, "large-footed."
Description Carpophore 3–4 cm in diameter, cup-shaped, ash-gray, granular-floccose outside. Stipe long, tapering slightly toward top, solid, sometimes sulcate at base, completely covered with gray wooly hairs that join to form tiny tufts. Flesh white, thin, waxlike. No odor or flavor. Spores white, elliptical, smooth, 20–30 × 10–12 microns.
Edibility Fair.
Habitat On the ground beneath broadleaf and coniferous trees.
Season Summer and autumn.
Note Other cup-shaped species also have a well-developed stipe. They belong to the genera *Cyathipodia; Helvella* (in the broad sense such as *H. queletii* or *H. cupuliformis,* with a light brown hymenium, grayish and hairy on the outside); and *Sowerbyella* (with the single species *S. radiculata,* autumnal, growing beneath conifers, also under elder shrubs, hymenium lemon-yellow, sterile surface wooly and cream-colored, with a long irregular stipe covered with dense white hairs).

418 RHYTISMA ACERINUM

Common name Tar-spot of maple.
Etymology From Latin, "pertaining to maple," because of the habitat.
Description Stromata up to 2 cm in diameter, fairly circular, distinct, shiny black, in the shape of barely raised plaques or scales on the leaf surface, containing numerous very small fruit-bodies (apothecia) which open outwards cracking the crust of the stroma, fertile surface gray. Spores colorless, filiform, smooth, 60–80 × 1.5–2.5 microns.
Edibility Nil.
Habitat On the leaves of sycamore maple, less common on other species of maple.
Season Summer, on living leaves; late winter through spring, on fallen leaves.
Note In the asexual state this fungus is called *Melasmia acerina,* and is parasitic on the living leaves of various maples. The sexual state with fruit-bodies develops inside the stromata on fallen leaves over winter and spring. *R. salicinum* is a similar fungus which grows on willow leaves.

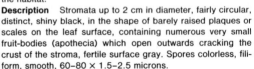

419   SPHAEROTHECA PANNOSA

Common name   Rose mildew.
Etymology   From Latin, "tattered," because of its appearance on the host plant.
Description   A very common mold or mildew in the Northern Hemisphere, it attacks roses and, in a similar form, peaches and apricots. The damage caused by this disease consists in a bubblelike deformation of the leaf laminae which are covered with a whitish powder, then turn yellow and wither. The buds and young shoots as well as the flowers are affected. When the petals are affected these become marked and dry out quickly, withering at the same time. The fruit are attacked in the case of peach and apricot infection. When mature the perithecia look like small black dots with a diameter of about 0.1 mm. Their surface is mosaiclike and they contain a single ascus. The fungus spends the winter in mycelial form (white).

420   USTILAGO MAYDIS

Common name   Corn smut.
Etymology   After its host plant, *Zea mays.*
Description   Parasitic fungus, living on maize, originally an American plant, now found in every country where cereal crops are grown. It attacks the aerial parts of the plant, and occasionally the root apparatus, producing irregular swellings or tumors which can reach a diameter of 15–20 cm. Those which form on the female inflorescence (the corn cob) are particularly large. Initially the tumorlike formation has a whitish coloration that gradually turns wine-red and then deep black. When mature the surface is papyruslike and fragile while the completely black inside is formed by a powdery spore mass. If these spores come into contact with food they are reported to cause disorders similar to ergotism (see **note** at *Claviceps purpurea*, **346**). The young tumors, on the other hand, which have a sugary and aromatic white pulp, are fairly sought after by some, and canned and sold commercially in Mexico, where they are considered a delicacy.

GLOSSARY

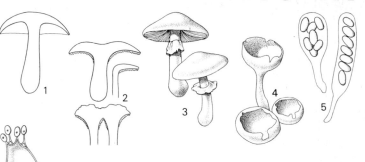

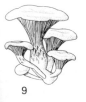

adnate (of gills) broadly attached to stipe (1).

adnexed (of gills) narrowly attached to stipe.

alveolate pitted like a honeycomb.

amyloid (of spores) when spores turn blue to blue-black with starchlike, iodine-based reagent (Melzer's reagent).

anastomosed (of gills and ridges) an angular network formed by cross-veins.

annulus ringlike structure on stipe, derived from partial veil.

apical positioned at apex or top.

apiculus (of basidiospores) short projection at or near base of spore where it was attached to sterigma of basidium.

apothecium cup-shaped fruit body, typical of certain Ascomycetes (4).

appendiculate (of cap margin) adorned with fragments of velar material.

areolate (of cap surface) cracked in quiltlike pattern.

ascus saclike microscopic structure which contains the spores of the Ascomycetes (5).

Ascomycetes fungi characterized by having asci, plural of ascus.

ball and socket (of cap and stipe attachment) where they are clearly separable from one another (18) (see *free*).

basidium club-shaped microscopic structure that bears the spores of the Basidiomycetes (6).

Basidiomycetes fungi characterized by having basidia, plural of basidium.

bulbous referring to an enlarged, bulb-shaped structure (7).

campanulate bell-shaped (8).

carpophore the fruit body or conspicuous and familiar part of the higher fungi, which contains the reproductive structures (asci or basidia).

cespitose tufted.

ciliate with hairs, like an eyelash (10).

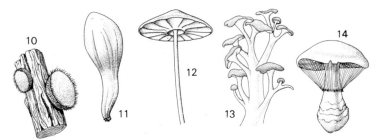

10 11 12 13 14

claviform   club-shaped (11).
collarium   the ringlike collar at the apex of the stipe that separates gills from stipe in certain fungi (12).
concrescent   growing together (13).
coprophilous   growing on dung or droppings.
cortina   cobweblike partial veil between cap margin and stipe of certain gilled fungi (14).
cuticle   outermost layer of cap or stipe.
cystidia   large sterile cells on cap, gill, or stipe surface.
decurrent   gills that descend stipe to some degree (15).

15

deliquescent   tissue that liquefies, like *Coprinus* gills.
dextrinoid   (of spores) when spores turn reddish brown with starchlike, iodine-based reagent (Melzer's reagent).
disc   central part of cap (16).
discomycetes   cup fungi, with asci exposed on apothecium.
eccentric   (of stipe) not attached in center of cap; off-center.

16

echinulate   covered with small spines or warts (17).
fasciculate   fungi with stipes in a bundle (19).
fistular   tubular, hollow (20).
floccose   cottony to downy-woolly.
free   (of gills) not attached to stipe (25).
fruit body   the carpophore or conspicuous and familiar part of the higher fungi that contains the reproductive structures (asci or basidia).
fugacious   soon disappearing.
fusiform   spindle-shaped (22).

17

gills   radially arranged platelike structures on the undersides of the caps of gilled fungi, which bear the basidia.
glabrous   without any hairs or other ornamentation.
hypha   a fungal filament that with others forms the hyphae of which the mycelium and flesh of the carpophore are composed.
hymenium   the spore-bearing surface of an Ascomycete or Basidiomycete.

18

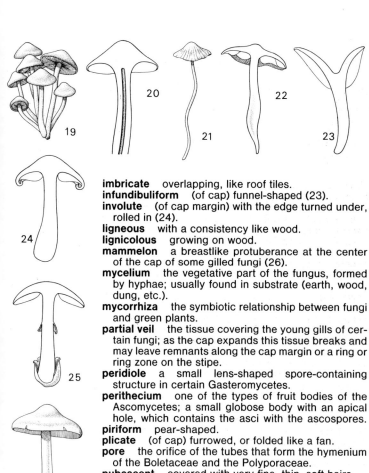

imbricate   overlapping, like roof tiles.

infundibuliform   (of cap) funnel-shaped (23).

involute   (of cap margin) with the edge turned under, rolled in (24).

ligneous   with a consistency like wood.

lignicolous   growing on wood.

mammelon   a breastlike protuberance at the center of the cap of some gilled fungi (26).

mycelium   the vegetative part of the fungus, formed by hyphae; usually found in substrate (earth, wood, dung, etc.).

mycorrhiza   the symbiotic relationship between fungi and green plants.

partial veil   the tissue covering the young gills of certain fungi; as the cap expands this tissue breaks and may leave remnants along the cap margin or a ring or ring zone on the stipe.

peridiole   a small lens-shaped spore-containing structure in certain Gasteromycetes.

perithecium   one of the types of fruit bodies of the Ascomycetes; a small globose body with an apical hole, which contains the asci with the ascospores.

piriform   pear-shaped.

plicate   (of cap) furrowed, or folded like a fan.

pore   the orifice of the tubes that form the hymenium of the Boletaceae and the Polyporaceae.

pubescent   covered with very fine, thin, soft hairs.

radicant   of the stipe when it penetrates the ground like a root (28).

resupinate   of a carpophore when it adheres to the substratum with its back or top (29).

reticulate   in the form of a net.

rhizomorph   mycelial formations resembling rootlike structures (30).

ring   the remains of the partial veil on the stipe of certain gill and pore fungi (3).

rugose   wrinkled, rough.

scabrous   rough, with short rigid projections.

sclerotium   a mass of tightly knotted hyphae forming very hard spheroidal or elongated structures. This

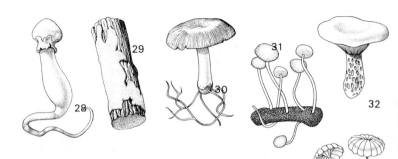

enables certain fungi to survive adverse environmental conditions (31).

scrobiculate having shallow grooves or depressions (32).

septate divided by cross-walls.

sessile without a stipe.

setaceous (of stipe) bristlelike (33).

squamose having flat scales on the cap (35).

stipe the part of the carpophore that supports the cap or the hymenium in general; the stem or stalk of the fungus.

striate marked with thin lines that are radial on a cap surface and longitudinal on a spore.

stroma (pl. *stromata*) compact mass of tissue on or in which perithecia are produced.

suberose eroded.

sulcate grooved, furrowed (34).

teeth small pointed structures found in the hymenium of the tooth fungi and certain polypores, on the surface of which the basidia are formed (2).

terrestrial growing on the ground.

tomentose (of cap or stipe) covered with wool or down.

tube small structure containing the basidia in the boletes and polypores.

tufted growing in tight groups (9).

umbo a central broad swelling, as on the caps of many gilled fungi.

undulate wavy (21).

universal veil the tissue covering the immature carpophore of certain fungi; on expansion of the fungus this veil breaks and leaves patches or remnants on the cap or a saclike cup or remnants about the base of the stipe.

velar pertaining to the veil (universal or partial).

verrucose having wartlike protuberances.

volva the remains of the universal veil which stay at the base of certain fungi (36).

BIBLIOGRAPHY

Ainsworth, G. C., Sparrow, F. K. and Sussman, A. S. (Eds.) *The Fungi: An Advanced Treatise,* Vols. IV A & IV B (A Taxonomic Review with Keys). New York: Academic Press, 1973.

Alexopoulos, C. J. and Mims, C. W. *Introductory Mycology,* 3rd edition. New York: John Wiley & Sons, 1979.

Guzman, G. *Identificación de los Hongos.* Mexico: Editorial Limusa, 1977.

Hesler, L. R. and Smith, A. H. *North American Species of Hygrophorus.* Knoxville: The University of Tennessee Press, 1963.

Hesler, L. R. and Smith, A. H. *North American Species of Lactarius.* Knoxville: The University of Tennessee Press, 1979.

Jenkins, D. T. *A Taxonomic and Nomenclatural Study of the Genus Amanita Section Amanita for North America.* Vaduz: J. Cramer Press, 1977.

Kauffman, C. H. *The Gilled Mushrooms (Agaricaceae) of Michigan and the Great Lakes Region.* New York: Dover Publications, 1971.

Largent, D., Johnson, D. and Watling, R. (consultant). *How to Identify Mushrooms to Genus III: Microscopic Features.* Eureka, California: Mad River Press, Inc., 1979.

Lincoff, G., and Mitchel, D. H. *Toxic and Hallucinogenic Mushroom Poisoning: A Handbook for Physicians and Mushroom Hunters.* New York: Van Nostrand Reinhold Co., 1977.

Miller, O. K. *Mushrooms of North America.* New York: E. P. Dutton, 1977.

Miller, O. K., and Farr, D. F. *An Index of the Common Fungi of North America (Synonymy and Common Names).* Vaduz: J. Cramer Press, 1975.

Moser, M, *Die Röhrlinge und Blätterpilze (Polyporales, Boletales, Agaricales, Russulales).* Stuttgart: Gustav Fischer Verlag, 1978.

Overholts, L. O. *The Polyporaceae of the United States, Alaska, and Canada.* Ann Arbor: The University of Michigan Press, 1967.

Pomerleau, René. *Flore des Champignons au Québec.* Montreal: Les Editions La Presse, 1980.

Ramsbottom, J. *Mushrooms and Toadstools.* London: Collins, 1954.

Singer, R. *The Agaricales in Modern Taxonomy,* 3rd Edition. Vaduz: J. Cramer Press, 1975.

Smith, A. H. *North American Species of Mycena.* Ann Arbor: The University of Michigan Press, 1947.

Smith, A. H., and Hesler, L. R. *The North American Species of Pholiota.* Monticello, New York: Lubrecht and Cramer, 1968.

Smith, A. H., and Smith, H. V. *How to Know the Non-Gilled Fleshy Fungi.* Dubuque: Wm. C. Brown Co., 1973.

Smith, A. H., Smith, H. V. and Weber, N. S. *How to Know the Gilled Mushrooms.* Dubuque: Wm. C. Brown Co., 1979.

Smith, A. H., and Thiers, H. D. *The Boletes of Michigan.* Ann Arbor: The University of Michigan Press, 1971.

Snell, W. H., and Dick, E. A. *The Boleti.* Vaduz: J. Cramer Press, 1970.

Snell, W. H., and Dick, E. A. *A Glossary of Mycology.* Cambridge: Harvard University Press, 1971.

Stamets, P. *Psilocybe Mushrooms and Their Allies.* Seattle: Homestead Book Co., 1978.

Thiers, H. D. *California Mushrooms: A Field Guide to the Boletes.* New York: Hafner Press, 1975.

Wasson, R. G. *The Wondrous Mushroom: Mycolatry in Mesoamerica.* New York: McGraw-Hill, 1980.

Watling, R., and Watling, A. E. *A Literature Guide for Identifying Mushrooms.* Eureka, California: Mad River Press, Inc., 1980.

INDEX

Note: Those numbers in Italics refer to the entry number; those in Roman refer to the page number of the Introduction.

Photographic References

Introduction
Ardea Photographics, London: 8–9; Lindau: 20.—Chaumeton, Paris: Chamalières: 50s, 50d; Lindau: 17, 34.—Coleman, London: 52, 53.—Jacana, Paris: 2.—Marka Graphic, Milan: 8, 9.—Natural History Photographic Agency, London: Preston Mafham: 12; Bain: 372, 373; Hawkes: 27, 32; Janes: 26, 31. Giovanni Pacioni, l'Aquila: 15b, 23, 37, 39a, 40, 47a, 47b, 48, 49.—Luisa Ricciarini, Milan: Bertola: 19; P2: 39b.

Entries

Alauda, Milan: Galli: 171, 210, 211, 216; Sessi: 212; Soresina: 201; Ardea Photographics, London: 228, 264; Usidan: 8, 87, 103.—Carrese, Milan: Serafin: 36, 40, 141, 200.—Bruno Cetto, Trento: 13, 14, 98, 103, 109, 115, 129, 135, 138, 139, 161, 169, 174, 178, 195, 221, 238, 263, 299, 337, 339, 377, 390, 417.—Chaumeton, Paris: 19, 33, 34, 35, 43, 55, 56, 59, 62, 64, 67, 71, 74, 81, 99, 102, 111, 116, 117, 118, 133, 152, 160, 184, 185, 186, 208, 226, 233, 252, 256, 273, 279, 281, 284, 286, 290, 302, 316, 317, 320, 370, 398; Lanceau: 1, 4, 5, 6, 7, 10, 15, 18, 21.—Coleman, London: 2, 31, 67, 75, 82, 89, 90, 92, 106, 121, 140, 170, 180, 191, 204, 230, 267, 274, 293, 296, 305, 306, 318, 325, 328, 329, 340, 343, 344, 345, 346, 347, 353, 356, 362, 365, 384, 394, 402, 410, 411, 412, 416; Bisserani: 360, 383, 385.—Florestano Ferri, 260.—Gruppo Micologico DLF, Verona: 16, 37, 48, 49, 58, 65, 155, 193, 206, 207, 237, 262, 376, 396.—Marcello Intini, Florence: 312.—Jacana, Paris: 11, 20, 23, 28, 29, 30, 63, 68, 91, 94, 95, 100, 114, 119, 205, 214, 248, 253, 261, 277, 295, 297, 300, 308, 330, 342, 354, 355, 357, 361, 367, 368, 369, 371, 373, 389, 405, 415;—Champoroux: 167; Hawkes: 166; Konig: 406; Nardin: 96, 151; Pilloud: 321, 399; Ruffier-Lanche: 278.—Marka Graphic, Milan: 335.—Natural History Photographic Agency, London: Bain: 107; Allen: 225, 231; Hawkes: 79, 80, 190, 229, 235, 239, 240, 254, 275, 387; Hyde: 9, 32, 45, 147, 198, 202, 223, 307, 309, 341, 358, 419; IDA: 150; Janes: 326; Preston-Mafham: 24, 25, 27, 38, 39, 42, 70, 84, 108, 131, 142, 145, 188, 222, 285, 334, 386, 400, 403.—Giovanni Pacioni, l'Aquila: 47, 50, 54, 60, 69, 77, 86, 97, 120, 122, 123, 132, 134, 148, 153, 157, 163, 169, 172, 199, 209, 213, 215, 217, 219, 227, 232, 234, 255, 259, 276, 280, 287, 289, 315, 319, 322, 323, 324, 336, 338, 348, 351, 352, 366, 378, 380, 382, 388, 391, 392, 401, 404, 407, 408, 414.—Luisa Ricciarini, Milan: 17, 44, 149, 162, 247, 249, 251, 288, 310, 311; Bertola: 243; Leonardi: 72, 78, 83, 104, 110, 127, 137, 144, 173, 175, 177, 194, 241, 250, 270, 271, 294, 359; P2: 16, 22, 52, 85, 124, 130, 182, 265, 372, 379; Shaeff: 331; Unedi: 46.—SEF, Turin: 113, 242.

The illustrations at the beginning of each chapter are by: Chaumeton, Paris (mushrooms with scales, pp. 70–71); Chaumeton, Paris (mushrooms with pores, pp. 320–321); National History Photographic Agency, London: Bain: (shelf or crust mushrooms, pp. 372–373); Jacana, Paris (club mushrooms, pp. 410–411); Coleman, London (spheres, stars, pears, and cup-shaped mushrooms, pp. 440–441); Marka Graphic, Milan (Glossary, p. 500).